THE
CONCEPT
OF
PROBABILITY

BY

J. R. LUCAS

Fellow of Merton College, Oxford

CLARENDON PRESS · OXFORD

1970

Oxford University Press, Ely House, London W.1

GLASGOW NEW YORK TORONTO MELBOURNE WELLINGTON
CAPE TOWN SALISBURY IBADAN NAIROBI DAR ES SALAAM LUSAKA
ADDIS ABABA BOMBAY CALCUTTA MADRAS KARACHI LAHORE DACCA
KUALA LUMPUR SINGAPORE HONG KONG TOKYO

MADE AND PRINTED IN GREAT BRITAIN BY
WILLIAM CLOWES AND SONS, LIMITED
LONDON AND BECCLES

TO

J. M. L.

H. M. L.

D. J. L.

FOREWORD

THE main argument of this book comes in Chapters III, IV, and V. In Chapters III and IV I show why our concept of probability must have the syntax it does. In Chapter IV I argue also for the special *rôle* of propositional functions in elucidating so-called "Conditional Probabilities" which I interpret as rules for re-assigning probabilities under a change of universe of discourse. In Chapter V, after proving Bernoulli's Theorem, I attempt to show that De Moivre's use of it to argue inversely from observed fre-quencies to probabilities is not, as is commonly affirmed by philosophers, fallacious. None of these arguments is new. R. T. Cox in an article in *The American Journal of Physics* in 1946, and in his *Algebra of Probable Inference*, Baltimore, 1961, anticipated, in slightly different form, my derivation of the canonical forms of the rules for the traditional calculus of probabilities. The sup-porters of the Logical Relation theory and their precursors have made many of the points I make in Chapter IV. And, besides De Moivre, most ordinary men have always believed the burden of my contention in Chapter V.

My reason for arguing these points afresh is that together they give us a coherent and complete account of our concept of pro-bability; which, I believe, none of the existing theories—the Sub-jective theory, the Logical Relation theory, the Frequency theory, the Equiprobability theory, or any other—does. I think I can show why we have the concept of probability, why it has to have the grammar it does have, and how it is anchored in the rest of our conceptual structure. The merits of existing theories are merits of my theory too: but I think their demerits remain peculiar to them alone. Many philosophical confusions which inhibit our thought on questions involving probability are cleared up, and the rationale given for some arguments we often use but find hard to justify. Finally, some important metaphysical con-clusions are adduced.

This is a philosophical book. It contains no original research on the mathematical theory of probability. But I hope some mathematicians researching in that field may feel happier about the foundations of their discipline, if they read what I have to say. Being a philosopher, I aim to be as unsophisticated as possible. I base my argument on Bernoulli's Theorem, not Borel's, and discuss no distribution more complicated than a Normal one. My silence is not intended to belittle the achievements of the Twentieth Century. It is only that most of the philosophical questions were already implicit in the work of the three preceding centuries: and although measure theory and the strong laws of large numbers do have philosophical implications, we cannot assess them properly until we have settled our opinions about more elementary results. For the same reason I have skimped the statistics. Statistical arguments are ones which very much need philosophical examination and assessment. But I suspect that that will be a task for some other philosopher than me. If my elucidation of the underlying concept of probability and my application of it to some of the very simplest patterns of statistical inference provides another man with the tools for a full critique of this important but puzzling branch of human thought, I shall be well content.

CONTENTS

I

THE GUARDED GUIDE

Probability, said Bishop Butler,† is the guide of life; but it is a guarded guide. We use it when we want to make an affirmation, but are not quite sure; and this uncertainty infects our understanding of the concept itself. We often have occasion to deal with probabilities, but have great difficulty in saying what probabilities *are*. We regard all probabilistic arguments with suspicion. Not only are they often tricksy, and fertile of fallacies, but the concept itself seems dubious, and unless it can be properly explicated, and shown to be soundly based, all arguments about probabilities are vitiated, and all inferences drawn with its aid are invalid. It has been made more difficult to understand by being considered too abstractly. Abstract nouns are confusing. We do better to start with the everyday adjectives and adverbs, and only when we have grasped their significance ascend to the higher levels of abstraction. In this, as in most other respects, probability is like truth. "*In vino 'veritas'*", says Austin,‡ "but in a sober symposium only '*verum*' ", and we may explicate the force of the word 'true' by comparing it with phrases such as 'I know that', 'I promise that', 'Trust me that . . .".§ When I say that I am telling the truth, I am plighting my troth to my hearer that he can trust what I am saying. "This is a true saying", says St. Paul,¶ "and worthy of all men to be received", and we could well take that as a

† Joseph Butler, *The Analogy of Religion*, London, 1736, Introduction, Section 3, *fin.*

‡ J. L. Austin, "Truth", *Proceedings of the Aristotelian Society*, Supplementary Volume, XXIV, 1950, p. 111; reprinted in J. L. Austin, *Philosophical Papers*, Oxford, 1961, p. 85; and in George Pitcher, ed., *Truth*, Englewood-Cliffs, 1964, p. 18.

§ Compare J. L. Austin, "Other Minds", *Proceedings of the Aristotelian Society*, Supplementary Volume, XX, 1946, pp. 170–2; reprinted in A. G. N. Flew, ed., *Logic and Language*, 2nd series, Oxford, 1953, pp. 143–4, and in J. L. Austin, *Philosophical Papers*, Oxford, 1961, pp. 66–8.

¶ I Timothy 1.15 (Prayer Book version).

definition. When we say that something is true we mean that it is worthy to be believed, and we are vouching for it and guaranteeing it. We are implying that there are good reasons for believing it, which we have been over and found cogent; and if, nevertheless, it turns out to be false, we shall be answerable for having led our listeners to rely on it.†

In contrast to the word 'true', the words 'probable' and 'probably' disclaim warranty.‡ They are like the words 'I think' or 'perhaps'. Just as I can guarantee that the train goes at 10.23 or that Peter will come to the party by saying 'I know that the train goes at 10.23' or 'I know that Peter will come to the party', so I can hedge by saying 'I think that the train goes at 10.23' or 'I think that Peter will come to the party'. If I say this, I give you some guidance, but I warn you not to rely on it. If you want to be sure of catching the train, or if you want to make firm plans, then you should look up a time-table or ring up the station to confirm the time the train goes. I have given you what help I can—it would have been churlish of me to remain silent, when asked what time the train went—but I warn you that my information is not reliable. It would be wrong to say 'I think the train goes at 10.23' if I had no reason at all for thinking the train went at 10.23 —if I had no idea at all, I should keep silence or say that I had no idea. It would also be wrong to lead you to place too much reliance on something which I may have misremembered, or which may have been altered since I last looked it up. Silence is too negative, assertion too positive. We need something in between. And this is probability. Instead of 'It is true . . .' we say 'It is probable . . .', instead of bare, unqualified assertion, we add the adverb 'probably'. In the same way, if Peter accepts an invitation but says he may have to cancel at the last minute as his wife is ill, we should tell other people that he will probably be there; we have reason to believe that he will be there—he has accepted the invitation— but we do not want our hearer to count on it, and so we guard against it by putting in the word 'probably'.

Thus we use the words 'probable', 'probably' and the phrases 'I think', 'in my opinion' *etc.* to give a tentative judgement. There is some reason, but not conclusive reason, for what we

† But see further, Ch. II, pp. 16–17.
‡ I am much indebted here to S. E. Toulmin, *The Uses of Argument*, Cambridge, 1958, Ch. II.

opine: but, *although not equally*, there is some counter-reason, which weighs the other way. It is a *dialectical* concept. There has been some debate—although usually an internal one—about what answer to give. "What time does the train go?" "10.23", we start to answer: we have some reason to think so—two years ago we went on it, and only a month ago, a colleague excused himself at 9.50 in order to catch a train; *but* they may have brought the summer timetables in, and have all the trains running five minutes early; moreover they have recently replaced the steam engines with diesels, and have withdrawn a good many trains completely: so we are not sure. Therefore we say 'probably'. It qualifies the 'therefore' that follows a 'but'.

The word 'probably' reports the conclusion of a debate in which there is something to be said on either side. We have reached a decision, but the argument was not all one way, and although the balance of argument is, we believe, in favour of the judgement we are giving, there was something to be said on the other side, and therefore our guidance is only guarded. We hedge. We do not give our opinion with full assurance or warranty. The word 'probably' acts as an escape clause. If the train does not go at 10.23 or Peter does not come to the party, we were not in the wrong. We have not misled any one, since we warned him from the outset that it was on the cards that the train would not go then, or that Peter would not turn up. The point of using the word 'probably' was that if, after having said that Peter probably would be at the party, we later found that he was not there, we should not have told a lie or unwittingly said anything untrue. If we had said simply that he would be at the party, and in fact he was not, then we should have to admit that we had been wrong in saying that he was coming: and if we had given a stronger assurance than merely saying that he was coming, if for instance we had said that we *knew* that he was coming, or had promised that he would be there, then we should have to eat our words if he failed to turn up, and would have to apologize to anybody who had taken our word for it that he would be there and had planned accordingly—it would be our responsibility to find transport for someone who had come relying on Peter's being there and giving him a lift home: whereas if we had used the word 'probably' we should have warned him not to count on it and to make alternative plans for getting home.

'Probably' is therefore a modal word. It affects not the content of what is said, but the way in which it is said, the force with which it is said, and the degree to which it is to be relied on. It avoids a head-on clash if the outcome is other than what we said it was probably going to be. There is no impropriety in saying first 'probably A' and later 'not A': and even after we have said 'not A' we can—sometimes—still maintain we were right to have said originally 'probably A': we were right to say the guest probably would come, even though in fact he did not. For, after all, he had accepted the invitation, and it would have been wrong to say that he was not going to come. Thus the addition of the word 'probably' to the statement 'He will be at the party' removes the flat contradiction which would otherwise hold between that statement and the subsequent admission 'He was not at the party'.

It is clear that although the chief function of the word 'probably' is to be a cushion against flat contradiction, this cannot be its only force. For if its only effect was to remove the risk of ever being wrong, it would make every sentence to which it was added meaningless. If there was no incompatibility at all between probably A and any other statement then the words 'probably A' would not interlock with any other part of our describable experience, and would have no place in our conceptual system.† To be effective, statements containing the word 'probably' must have some rules for its use, some criteria for its correct and incorrect application. And of course it has. If I say 'He will probably come to the party', it is not enough that I do not want to commit myself to justifying my statement. To be justified in saying 'He will probably come to the party' I must have some reason for believing that he will be coming; or at the least that I am not uttering this entirely arbitrarily. I should be justified if I could give some reason—e.g. that he had accepted the invitation, albeit with a proviso—or perhaps even if I only had a hunch, which I was willing to back though unable to exhibit a cogent argument: but I should not be justified in saying 'He will probably come to the party' if I had no grounds whatever for thinking that there would, and was saying it quite idly but taking care to guard against my being held to account for it when in fact he did not come. The word 'probably' has a gerundive force, like the word

† See, more fully, J. P. Day, *Inductive Probability*, London, 1961, pp. 294–5. See also below, Ch. V, p. 91.

'true'. Although a guarded guide, it is none the less a guide. When I say 'probably A', I mean not only that I think that A and want you to believe it too, but that there are *reasons* why I believe A and why you *should* believe it. Only, unlike the word 'true', the word 'probably' does not imply that the reasons are all that one could want. Although there are reasons, there is room for reasonable doubt too. So we say what the reasons would lead us to conclude, but indicate at the same time that there is a real possibility of its being otherwise and write in an escape clause.

In the example we have been considering there is a specific contingency we have in mind when we hedge. We have good grounds for believing that Peter will come to the party, but he will not come if his wife is ill. To any one who expresses disappointment at not seeing him at the party we will say that it had only been probable that he would come, and in fact his wife had not recovered sufficiently after all. More formally, we use the word 'probably' and say probably A in the sense we are considering now, when we know certain circumstances, B, C, D, which together are sufficient, granted that all other conditions were normal, for us to be able to infer A, but we have reason to believe that some of the other conditions may not be normal, in which case it will not come about that A is true. We therefore say probably A, having reason to believe that A will prove to be true, but already having to hand an explanation of not A, should not A turn out to be the case.

We need not, however, have a specific contingency in mind. We can quite properly hedge when there are a large number of standing conditions *sine qua non*, any of which could fail and could disappoint our expectations: there is many a slip between cup and lip, we may say, hedging on complicated predictions requiring many assumptions which may not in the event hold good. We have reason for believing that A will prove to be true, but if it should turn out otherwise we have some idea of the sort of explanation we shall offer.

The concept of probability considered thus far has its habitat among those concepts of assertion and counter-assertion, objection and rebuttal which occur naturally among our unformalised modes of *argument*.† We are giving the conclusion of an argument, which

† For a general discussion of these concepts, see H. L. A. Hart, "The Ascription of Responsibility and Rights", *Proceedings of the Aristotelian*

on balance goes in favour of an original unqualified assertion—
"He will be at the party", but conceding the strength of the objection. We are putting forward an already partially rebutted
assertion, ready to be completely falsified by the event, though
not expected to be. 'Probably' means 'arguably'—with the implication 'arguably, but', 'it is arguable that . . . , but on the
other hand . . .'. By a curious reversal of fortune the word has
lost caste in the course of its history: etymologically it is derived
from the Latin '*probare*' 'to prove', and in the middle ages it had
the sense of 'provable', implying certainty. In the sixteenth century it was used sometimes to mean only 'credible' or 'plausible',
and since then it has lost all its implications of certainty, and acquired just the reverse. 'Provable' has come to mean 'arguable'
and 'arguable' has become 'arguable, but'.

Like the other modal words 'possibly' and 'necessarily', 'probably' varies its force with its context. If I say "You cannot possibly go to the party without a tie", I am speaking of what is
socially possible or permissible, not of what is legally, morally,
biologically, physically or mathematically possible. In the same
way with 'probably', the context will show the sort of certainty I am
being careful not to guarantee. If I say "My brother will probably
be calling this afternoon", because he usually does look round some
time before tea, I am hedging against the sort of certainty that I
should have had if he had told me that morning that he definitely
would look in. But even if he had told me, I might still want to
hedge against the possibility of his changing his mind, and so
again would use the word 'probably', but this time with a different
force, because a different sort of assurance was being excluded.
And even when an entirely trustworthy person has promised, we
still may need to guard against accident or sudden illness. All
human actions are subject to the changes and chances of this
fleeting life, and are to that extent less than certain. By the same
token, medical and biological certainties are less than absolutely
certain—incurable cancers have been known to disperse, men

Society, XLIX, 1948, pp. 171–94; reprinted in A. G. N. Flew, ed., *Logic
and Language*, 1st series, Oxford, 1951; see also, J. R. Lucas, "The Philosophy of the Reasonable Man", *The Philosophical Quarterly*, 13, 1963, pp.
97–106; "The Lesbian Rule", *Philosophy*, XXX, 1955, pp. 195–213; and
"Not 'Therefore' but 'But'", *The Philosophical Quarterly*, 16, 1966, pp.
289–307.

given up for dead have come to life in their coffins. Even the laws
of chemistry and physics do not have that peculiarly ineluctable
certainty that mathematical and logical truths have. And we may
mark the distinctions in each case by saying that biological
generalisations are only probably true, as compared with those of
chemistry, or that the generalisations of physics are only probably true, as compared with those of mathematics. Thus Locke,†
and more recently the logical positivists, were driven to deny that
we really knew things which we normally would say that we did
know: every empirical generalisation is "but probability, not
knowledge" because "I have not that certainty of it which we
strictly call knowledge".

We thus need to be continually considering what the counter-
argument is which leads a speaker to hedge. It is logically possible that the sun should not rise tomorrow, that water should be
no longer wet, and gold at ordinary temperatures run like mer-
cury. We cannot rule out such contingencies by logic alone. And
therefore, say Locke and Ayer, we cannot be as certain about
them as we can be that all bachelors are unmarried, that either it
is raining or it is not, and that two and two make four. Synthetic
generalisations are only probable because they fail to be analytic
truths. We may be very confident that the sun will rise tomorrow:
but, urge Locke and Ayer, we cannot be as confident about that
as we can be that a bachelor is unmarried: for the contrary supposition, that the sun will not rise tomorrow is at least intelligible,
whereas the suggestion that a bachelor might be married can be
dismissed at once as nonsense. Anyone who says a bachelor could
be married, or that it might be neither raining nor not raining, or
that two and two make five, thereby shows he does not understand the meaning of the words he uses. But the man who says
that the sun will not rise ever again after having set tonight cannot be ruled out of court by the mere statement of his claim. And
this 'but' they express by saying that empirical generalisations
are only probable and cannot be known for certain.

We may protest that Locke and Ayer are using the words
'knowledge', 'probability', and 'certainty' in an unusual way:
but that does not dispose of their contention, as some linguistic
philosophers have supposed. Rather, we should accept the point

† John Locke, *Essay Concerning Human Understanding*, Bk. IV, Ch. 6,
§16; Ch. 11, §9.

that is being made, but go on to say that the words 'probably' 'certain' *etc.* are "systematically ambiguous", and change their exact colouring from context to context; so that although it is true to say that compared with mathematical statements it is only probably true that the sun will rise tomorrow, compared with statements made in a court of law it is certain that the sun will rise tomorrow. We can, if we wish, use the words 'probably' and 'certain' to call attention to the fact that the laws of physics are not susceptible of deductive proof from logically necessary premisses, or that biological principles are not like physical laws, or that generalisations about human affairs are not scientific statements, or that many judgements about historical events are not so firmly established that they could not be upset by further evidence. But we must not assume that exactly the same distinction is being drawn in every case, or that what is said to be merely probable in each case is of the same type as in every other case. Much confusion has been caused by this, often unrecognised, assumption. Writers have felt that they should be giving, all in one breath, an account of chance events, of the justification of inductive arguments, and of rules for historical and legal reasoning, because each is in some sense a matter of probabilities rather than of certainties. One might as well include a chapter on social etiquette in a work on modal logic. All in fact we need to do is to take care to distinguish the different senses of the word 'probably', just as we ought to be sure, whenever we use the words 'possibly' or 'necessarily', what sort of possibility or necessity—logical, physical, linguistic, moral, social, *etc.*—we mean. In our case we must consider what sort of certainty we are intending to exclude: when we say that it is probable that smoking causes cancer, we may be saying that while the figures show a connexion, we do not yet know what the causal mechanism is; hence the hedge. Or we may be saying that cigarette smoking is not the only factor correlated with lung cancer—that living in towns is also a factor, and we do not know properly the significance of each separately. Or we may be saying that the figures can be explained by some alternative hypothesis—that only people with a predisposition to cancer take up smoking. Or we may be saying that the figures are not statistically significant. Or that they have been cooked by the Ministry of Health and the Band of Hope in a secret conspiracy. Or we may simply be saying that there are a lot of smokers who live to a ripe old age and do not die of lung cancer. The words 'probably' and 'probable' can be used to leave room for any one of these counters. And so whenever a probability-judgement is made, we need to consider both the sort of grounds on which the judgement could be based and the nature of the 'but' for which provision has been made.

Dialogues are often difficult to engage in. Sometimes there is no opportunity of asking the speaker why he hedges his assertion; or there may not be time; or we may not want to be bothered. When a man says 'It is true that . . .' or 'He knows that . . .', although the use of these phrases indicates that he has good reasons for saying what he says which he will produce *if* we ask him, we normally do not ask him, and are content to accept his word for it, and concern ourselves only with what he says, not why he says it. The same is true of 'probably'. Although the concept is rooted in a dialogue, it is frequently used and considered in the context of a monologue, where no arguments and counter-arguments are communicated, but only conclusions. In its new habitat, the concept has grown more formal and sophisticated; but also more abstract and more easily misunderstood.

II

A GRADUATED GUIDE

THERE are many reasons for hedging, not all equally cogent. If we are not to give, and cannot be asked, our reasons in full, so that our hearer may evaluate them for himself, we need at least to enable him to compare our hedges, and know how guarded our guidance is. Else, all he would know would be that our statement was insufficiently reliable for us to guarantee it, and that it was consistent with any later statement about the same topic that any one might care to make. But if that were all he could know, then the concept would be an idle one, and useless. As we move from the context in which reasons can be asked for and given, so we come to require a probability-judgement to indicate on its face how guarded it is. At the very least, we must be prepared to say sometimes whether one conclusion is more guarded—that is, less probable—than another. Nor can we resist the question, and claim that it is impossible to compare different arguments. For it was only because there were arguments both for and against a conclusion, and the arguments for were *stronger than* those against, that we could assert anything, albeit only guardedly, at all. Therefore we must be able sometimes to compare arguments, and hence the guardedness of our guides.

If we can assess arguments on both sides, and say which way the balance of argument goes, we may be able to compare them in different cases, and say how much more the argument in favour outweighs the argument against in the one case than in the other. It may be only a matter of degree—we discriminate between some conclusions that are very probable from others where the antecedent arguments are fairly evenly balanced. In cases where the arguments for and the arguments against are reasonably comparable, we can compare one judgement with another more closely. It is more probable that the battle of Marathon was in 490 B.C. than that the battle of Carchemish was in 606 B.C. but more probable still that the battle of the Halys was in

585 B.C. But where the conclusions, or the arguments leading to them or bearing against them, are of different kinds, it is doubtful whether comparisons of probability can be usefully made. Is it more probable that the sun will rise tomorrow than that George VI died in 1952? Is the Theory of Evolution more probable than that a thousand tosses of a true die should not yield a thousand sixes? What sense are we to make of these questions? Just as the word 'true' applies in many different disciplines, without there being one uniform way whereby true propositions obtain their truth, so the word 'probably' applies in many different disciplines without there being one uniform scale on which all probabilities may be compared. In the language of the logicians, 'more probably than', although an ordering relation, does not establish a complete ordering.

If we can compare probability-judgements, even though only sometimes, we have a use for the phrases 'more probable than' and 'less probable than', and hence a very rough 'more or less as probable as'. We can back-form the positive adjectives 'probable' and 'improbable' from the comparatives 'more probable than' and 'less probable than', in the same way as we mean by 'tall', 'taller than average'.† In particular, we can construct, again very roughly, some notion of equiprobability to act as a standard by means of the locution 'as probable as not'. Comparisons of probabilities are often impossible, and always, thus far, crude and inexact: but the possibility of their being made is inherent in the dialectical root of the concept, and is what comes to the fore when the concept is transplanted into the context of a monologue. Instead of the modal adverb 'probably' indicating some sort of hedge, to be elucidated by subsequent question and answer, we have an adjective 'probable' or abstract noun 'probability' indicating degrees—degrees of something lying between truth and falsehood. It is not the only possible development. 'Probable' used to mean 'arguable', and it retained until recently a sense in which it could be applied not to propositions and the like, but to arguments, arguments which were less than conclusive but none the less carried some weight.

There has been much confusion about the relation between arguments and conclusions in probability theory. I shall attempt

† See below, Ch. IX, p. 167.

to clear them up later.† At present I only want to indicate a parting of the ways, and choose one of them. We may choose to concern ourselves with arguments, and seek to generalise the all-or-nothing concepts of validity, invalidity, *etc.* If so, we shall be led to construct some sort of confirmation theory, filling in the graduations between *three* extremes: arguments which are altogether valid or sound, where the premisses entirely establish their conclusion; arguments which are the reverse of this, where the premisses completely " disconfirm" or refute the putative conclusion; and arguments which do nothing either to prove or to disprove their conclusions, where the premisses are entirely irrelevant to the question at issue. I shall not follow this path. Although, as I have argued, we are able sometimes to compare the strength of arguments, and must be able to if we are ever to make any probability-judgements, I shall confine myself to what we actually say about the conclusions, and not attempt to elucidate what we might have said, if asked, about the arguments which led us to them. The concept of probability has emigrated from its native dialectical shores and taken out naturalisation papers in the land of unbroken monologues: it has come to occupy the same territory as truth and falsehood, and is engaged in the same line of business, filling in between them. The classical theory of probability is a theory of something which, like truth and falsehood, can be ascribed to propositions, propositional functions, statements, sentences, or judgements, but, in some way or other, lying between truth or falsehood. And it is this concept, or programme for a concept, that I shall elucidate and develop.

That probability was a matter of degrees was understood by the Cambridge thinkers, W. E. Johnson, Lord Keynes, and Sir Harold Jeffreys. They defined probability as a degree of rational belief.‡ It was an unfortunate definition in that it led other thinkers§ to suppose that, despite their explicit protestations to the contrary, ¶ they were subjectivists. The objective import of the word 'rational' did not effectively counter the subjective con-

---

† In Ch. IV.
‡ J. M. Keynes, *A Treatise on Probability*, London, 1921, pp. 4, 8, 11. H. Jeffreys, *Theory of Probability*, 3rd ed., Oxford, 1961, p. 30.
§ *e.g.* R. von Mises, *Probability, Statistics and Truth*, 2nd ed., London, 1957, pp. 75, 78.
¶ J. M. Keynes, *op. cit.*, p. 8.

notations of the word 'belief'. If we took the word 'belief' in the sense in which it means more or less the same as 'proposition'—as in "The belief that the earth is flat is untrue", to say that probability was a degree of the rational rightness of belief would be unobjectionable: but since the word 'belief' can be construed as something appertaining to persons, and varying from person to person—as in "It is his belief that the earth is flat, but not mine" —and since there are other powerful pressures to subjectivism, it is important to be explicitly non-subjectivist in any definition we adopt.

Subjectivism is a philosophical position easy to adopt, difficult to maintain consistently. We adopt it either because we doubt our own convictions or dislike our opponent's. "Who am I", the diffident young man asks, "to claim to know the truth? I may well be wrong. I cannot pretend to perfect wisdom. I am not entitled to say that my opinions are the true ones, but only that they are mine." "Who are you", the belligerent youth demands, "to lay down the law for me? Speak for yourself. You cannot tell me what to think. What you say has no relevance for me, no claim on my allegiance. It is not the truth, only the expression of your misguided opinion." Both arguments seem irresistible, and we become subjectivists. But it is difficult to remain occupied only by one's own opinions, uninfluenced by those of others. We want to communicate. And only under some ideal of non-subjectivity is communication possible. Only if it is logically possible for you to disagree with me is your agreement worth having or can I be telling you anything at all. If I am confined to making true statements about my state of mind and you about yours, we can never converse because we shall never be talking about the same topic. Our conversation will be like F. P. Ramsey's "I went to Granchester yesterday", "Oh, did you? I did not!" † But we are not thus separated logically into solipsistic cells. I can make remarks which, whether they are right or wrong, *claim* to be right, and which, whether you accept them or reject them, *call* for your acceptance. Indeed, so strong is our desire to communicate that we are prepared to construe as objective many statements which on the face of it are patently autobiographical and subjective. If I say 'I think that the train goes at 10.20', the grammatical form of the sentence would seem to show that it is a statement about my

† F. P. Ramsey, *The Foundations of Mathematics*, London, 1931, p. 289.

opining. But we do not understand it so. If my friend wants to disagree with me, he normally does not use the sentence 'You don't think that the train goes at 10.23' but says instead 'I don't think it does'. The subjective sense of 'I think' is so little used that the phrase has become, like the Latin *ut opinor*, and the Greek ὡς ἐγῷμαι, a "parenthetical verb",[†] whose function is to modify and weaken the force of the subordinate clause, not construct a new statement about it. 'Probably' and 'probable' are like 'I think'. If I say 'Peter will probably be at the party' and you say 'He probably won't' we are contradicting each other, although only guardedly. We are neither of us guaranteeing our assertions, but we are both making assertions, and incompatible ones, we are both offering guidance, and nobody could be guided by us both.[‡]

Of course we shall not leave it at that. We shall argue, and each will give his reasons for stating that Peter probably will or probably will not be at the party. At the end of it we may reach agreement or have to agree to disagree. If we reach agreement, it may well be because one, or both, of the parties to the argument has pointed out some facts not known to the other. Subjectivists and others have construed this as showing that probabilities attach not to conclusions by themselves but to conclusions-on-the-basis-of-evidence, so that, as the evidence changes the probability inevitably changes too. That the argument is not compelling can be seen by replacing 'probable' by 'true'. I may say 'It is true that the train goes at 10.23' and you say 'It is not true'; again, we shall argue, and may adduce facts which lead us to agree. But we do not therefore say that truth is relative to evidence. Although evidence is relevant to truth, what I am claiming to be true is simply the proposition put forward, that the train goes at 10.23. Relevant evidence may lead me to change my mind. But if it does, I am *changing* my mind. I am saying something different—namely that it is not true—about the same thing—namely that the train goes at 10.23—not the same thing about two different things. Similarly with probabilities. If, having said that it was probable that the train would go at 10.23, I am told of a new piece of information—*e.g.*, that the summer timetable has just come into force—I may revise my judgement, and say now that it is not so probable

† J. O. Urmson, "Parenthetical Verbs", *Mind*, LXI, 1952, pp. 480–96.
‡ See further below, Ch. IV, pp. 53–6.

as I had previously thought. But I shall then be saying something different about a train going at 10.23, not talking about a different topic. My hearer may not possess all the evidence in my possession. He may not know relative-to-what my probability-judgement has been made. All he knows for certain is what I have told him: and if I now say that it is not so probable that there is a train at 10.23 as I had previously said, he must construe this as giving him guidance about trains that is more guarded still; he must rely even less on there being a train then or thereabouts; it would not do to telephone the station after breakfast tomorrow, he must find out today, in case there is no suitable train after the 8.18. Probability cannot be relative to evidence which is often unspecified, any more than truth can: it must be ascribed to what has been explicitly stated. For only so can it be of any use.

Probability, however, is not identical with truth, and it might be that the differences between the two concepts were enough to make one, but not the other, a subjective one. Ordinary language sometimes seems to support the subjectivists. When we alter our assessments of probabilities, we are often reluctant to withdraw our original assignments. "It *was* probable that it would rain", we say, as the sun shines remorselessly from a cloudless sky, "although it turned out false". "We know now that the stories which Marco Polo told on his return to Venice were true, however improbable they may have been for his contemporaries."† But these locutions do not prove the case. As Toulmin argues,‡ they often obtain their force by slurring over the distinction between *oratio obliqua* and *oratio recta*, sometimes by using 'was probable' for 'seemed probable', sometimes by confusing the (good) grounds for the probability-judgement with the content, which, though hedged, was nonetheless falsified in the event. It was reasonable for the Venetians to disbelieve the stories of Marco Polo; they could legitimately use the word 'improbable' of them; and we too can use it vicariously in *oratio obliqua* on their behalf. Similarly, in self-justification, I can affirm, on my own past behalf, that the word 'probable' which was on my lips was then correctly there. But in neither case can I properly re-authenticate the word. I cannot say, straightly and without indication of oblique reference, that rain today is probable, or that Marco Polo's tales

† W. C. Kneale, *Probability and Induction*, Oxford, 1949, §1, p. 1.
‡ S. E. Toulmin, *The Uses of Argument*, Cambridge, 1958, pp. 55–62.

are improbable. If I were to maintain either of these propositions, I should be open to devastating counter-attacks, and I should be forced to withdraw. It is only when I am talking at a sufficient distance from my own or other people's erstwhile statements, that I can detach the word 'probable' from its gerundive force, and use it, as it were to exculpate my previous self's or other people's statements, which although wrong and not worthy of belief, were reasonably arrived at. Normally the word 'probable' tells one's hearer what it is reasonable to believe, but with the warning that one may be wrong: in *oratio obliqua*, however, and covered by the word 'was', its *rôle* is reversed, and it exculpates instead of hedges, saying that although the belief is wrong, it was a reasonable one to believe. Thus far, but only thus far, the subjectivists are right. The locution on which they take their stand is a real one, but not a central one. The words 'probable' and 'improbable' unlike the words 'true' and 'false' can be divested of their gerundive force and used in oblique contexts to stress how reasonable it was at the time of utterance to make the statement, in spite of its actually having been wrong to do so, instead of, as normally, being used to commend a belief, but with the proviso that it none the less could be wrong.

Toulmin draws an illuminating distinction between "Improper Claims" and "Mistaken Claims".† An improper claim is one which ought not to have been put forward because the claimant did not at the time have good grounds for it. A mistaken claim is one which is not borne out by events. If I make a guess, it would be improper to claim that I know: but, if it comes out right, I cannot be said to have made a mistake. Truth is more firmly tied to the non-mistakenness of the claim than to its propriety. Although I ought not to say that it is true that the train will go at 10.23 unless I have good reason for thinking so, when the crunch comes I stand or fall by the train's actually going or failing to go then, and not by my grounds for having believed it. In this one respect, to say something is true is different from saying that one knows it: for although one cannot know what is not true, in the sense that one has to withdraw the claim to knowledge in the event of having been mistaken, the fact that what one said turned out to be true does not prove that one knew it. Truth has become fully acclimatized to its monologous habitat: it always refers to the

† S. E. Toulmin, *The Uses of Argument*, Cambridge, 1958, pp. 57–62.

trustworthiness of what is said rather than of the man who said it. And therefore never loses its gerundive force.

Acclimatization is more difficult for probability, just because guarded statements cannot be brought up sharp by confrontations with awkward facts. The truth or falsity of the proposition 'the train goes at 10.23' can be decided definitely by the event: but the guarded assertions contained in probability-judgements are cushioned against being simply proved mistaken by events, and therefore it is more difficult to abandon the connexion with their propriety rather than their non-mistakenness, and in oblique and unusual contexts it reasserts itself. But it is still a secondary sense, as we can see if we compare the phrases 'was probable' (or 'improbable') and 'seemed probable' (or 'improbable'). There is no difference between saying "We know now that the stories which Marco Polo told on his return to Venice were true, however improbable they may have *been* for his contemporaries" and saying "We know now that the stories which Marco Polo told on his return to Venice were true, however improbable they may have *seemed* to his contemporaries". Indeed, under pressure, I think we should offer the latter as a more felicitous expression of what the former was trying to say. In oblique contexts there is no difference between 'was probable' (or 'improbable') and 'seemed probable' (or 'improbable'), and the former is only a *façon de parler* for the latter. In straight contexts, however, the distinction re-appears. We use 'seems probable' as a double hedge: not only might our conclusion be wrong—hence the 'probable'—but we may not have taken into account all the available evidence or given the question proper consideration. I start the argument with 'seems probable' and invite you to produce further evidence or adduce counter-arguments. When we have done all this, then we conclude with 'is probable': it may yet prove wrong, but not because we have overlooked any material consideration available to us now. We cannot say that Marco Polo's stories seemed improbable to the Venetians but were not really improbable. 'Were improbable' just means 'seemed improbable'. We can say of the Ionian philosophers, that their scientific speculations seemed probable to them but *are* not really very probable: or that the doctrine of continuous creation seemed probable to Professor Hoyle, but is not really so. In such cases we are not ruling out the possibility of our being wrong: everything may be made of fire or of water or of hydrogen nuclei, or the doctrine of continuous creation may be true after all; only, these are improbable beliefs, which we are, guardedly, recommending people not to hold; nevertheless, they seemed probable to those who propounded them. Thus 'was probable' (or 'improbable') means the same as 'seemed probable' (or 'improbable'), and is different from 'is probable' (or 'improbable'). The phrases 'improbable, but true' and 'probable, but false' do not show, as the subjectivists claim, that 'probable' and 'improbable' do not normally have a gerundive force, analogous to that of 'true' and 'false', only hedged; rather, the contrast is between what seemed to be and what really is, with the 'seem' expressed by the oblique context and past tense.

One counter-example needs to be considered.† If in a thousand tosses of a coin, exactly five hundred were heads and five hundred tails, the man of common sense would not be surprised. It would be exactly what he expected. A person acquainted with probability theory would not agree. Although it may seem probable that a thousand tosses should come down with five hundred heads and five hundred tails, in fact it is highly improbable that there should be exactly five hundred of each. Nevertheless, in this case, it is true. Hence it appears that we can have something seeming probable, but being improbable, and yet still being true. Therefore 'improbable, but true' is reinstated, and cannot be explained away as meaning really 'seemed improbable, but is really true'.

The answer is in two stages. First the distinction between seeming probable and being improbable is, untypically, a matter of more mature consideration rather than fuller information. The man of common sense had not calculated what the real probability of getting exactly five hundred heads and five hundred tails was: perhaps also he had not appreciated the force of 'exactly'. The antithesis here is between the unconsidered and considered statements of the probability of a certain complex propositional function (namely, that a sequence of a thousand tosses of a coin which has a probability of coming down heads equal to $\frac{1}{2}$ should yield exactly five hundred heads and five hundred tails). Secondly, however, we are faced with a distinction of a different sort. The contrast is between the assignment of a certain, low, probability to a *propositional function* (the one in brackets five to three lines above), and the

† Which I owe to Mr. D. F. Goda, of Merton College.

assignment of a certain truth-value—namely True—to a *proposition* (namely that *this* coin on *this* occasion has come down exactly five hundred heads and five hundred tails). That contrast is a typical one, and one we shall need to consider much more fully.†

The argument from *oratio obliqua* does not show that probabilities are relative to evidence, although it does show that the gerundive force of 'probable' is more easily muted than that of 'true'. Another argument, however, seems more telling. The probability of some predictions seems to change with time. It will probably rain tomorrow, I say, as the barometer falls, and white strands of cloud gather high in the sky. But after the clouds have cleared again and the barometer risen, rain tomorrow has become less probable. We might insist that with past tenses, as in indirect speech, 'probable' was elliptic for 'seemed probable'; but it would be a procrustean treatment. In fact, there is something else about singular propositions which explains the apparent variability of the probability we ascribe to them without making it a purely subjective assignment. But we must leave the discussion of this until Chapter VI.

One final objection remains. There is an instinctive repugnance to "objective" probabilities. Either Peter is in London or he is not. There is no other possibility, no half-way house, and if only the truth were known we should know which. To fill the world with absolute probabilities is not only unnecessary, but impossible. Things are as they are and cannot be only probably as they are, nor can they be probably not as they are. If we were less limited and ignorant, we should know how things were in the same way as God knows them, and then for us, as for Him, there would be no probability, only certainty.‡ Probability is thus seen

† In Ch. V.
‡ Compare Butler, *The Analogy of Religion*, London, 1736, Introduction, §3: "Probable Evidence, in its very Nature, affords but an imperfect kind of Information; and is to be considered as relative only to Beings of limited Capacities. For Nothing which is the possible object of Knowledge, whether past, present, or future, can be probable to an infinite Intelligence; since it cannot *but* be discerned absolutely, as it is in itself, certainly true or certainly false." Or W. A. Whitworth, *Choice and Chance*, 5th ed., Cambridge, 1901, Ch. VI, p. 121: "Thus the probability is seen to be entirely conditional on the respective degrees of our knowledge and ignorance; and so soon as our ignorance vanishes—so soon as we know all about the event, and become *as far as that event is concerned* omniscient—

to be only the measure of our ignorance and finitude, and to have no place in the real nature of things. The objection has considerable force. It may be weakened, however, by the *tu quoque* that a correspondence theory of truth leads to comparable infelicities. The objections we feel to reified probabilities hold equally against any theory that regards truth as a substantial entity: but truth can be objective without being reified; and so too, I maintain, probability. The justification for having a *concept* of objective probability is that we can use it. And provided we can give consistent rules for its use, it would be a wrong-headed use of commonsense metaphysics to ban its use on the grounds that probabilities do not exist in the outside world.

Even so, some difficulty remains, and we shall have to discuss it at length in Chapters XI and XII. Our conceptual structure is not totally divorced from our ontological commitments, and we feel unhappy about a concept which is, at some fundamental level of reality, always without application. Though they might be objective, for the reasons just given, probabilities would still be only a second-best, if they were always to be replaced in any full and definitive account of the world. We have a Laplacean view of God, and find it difficult to envisage a God's-eye view of the world which is not in all respects completely definite. Often our probability-judgements are made only because we have insufficient information, and it is natural to assume that this is always so, and that for God probability, like hope, vanishes into sight. But it need not be so. Perfect information may be unattainable even in principle; indeed, quantum mechanics seems to demand just this, and we may be compelled to believe that objective probabilities do exist, and cannot be accounted for simply in terms of our subjective lack of perfect information. If so, our view of the

---

then there no longer remains a question of probability; the probability is replaced by certainty." Or B. Finetti, "Foresight: its logical laws, its subjective sources", in H. E. Kyburg and H. E. Smokler, eds., *Studies in Subjective Probability*, New York, 1964, p. 147: "In the last analysis, each evaluation of probabilities different from 0 or 1 will surely be abandoned, for a well-determined event can only happen or not happen; an evaluation of probability only makes sense when and as long as an individual does not know the result of the envisaged event", quoted by D. H. Mellor, "Chance", *Supplementary Volume of the Proceedings of the Aristotelian Society*, XLIII, 1969, §1, p. 16.

world will be altered. To that extent we take the objector's point that our conceptual structure should reflect our ontological commitments, and *vice versa*. He cannot argue on grounds of logic that objective probabilities cannot exist, but he may maintain on metaphysical grounds that they do not exist. And if we, contrariwise, maintain that they do, we must be prepared to pay a considerable price in the alterations we shall have to make in our view of the world.†

The subjectivist account of probability, although natural, is wrong. It is natural, because the concept of probability has not developed an independent existence of its own to the same extent as that of truth, and is correspondingly more dependent on the justification the speaker had for using it: but it is wrong because in the end the concept is, like truth, a gerundive one, giving us guidance on what we ought to believe. Although in indirect speech and non-present tenses we fall back on the justifiability rather than the non-mistakenness of a probability-judgement, in the paradigm present indicative of direct speech 'It is probable that . . .' guides belief in the same way as, although more guardedly than, 'It is true that . . .'. This is the sense which is developed when we come to have a theory of probability, in which everything that is to be understood must be stated explicitly and nothing can be left to be elucidated by further question. In such conditions the word 'probable' must grow in the same way as the word 'true', and at a higher level of abstraction probabilities are entirely *pari passu* with truth and falsehood. Just as the simple underwriting- and denial- operators 'It is true that . . .' and 'It is false that . . .' give birth to the concepts of truth and falsehood which are the metalogical properties—or better, metalogical *values* (in this case, discrete "truth-values")—that can be ascribed to propositions, propositional functions, statements, sentences, formulae, or judgements, so the simple hedging operator 'It is probable that . . .' gives birth to the abstract noun 'probability', which in our sophisticated discourses we can ascribe various degrees of partial truth. There are philosophical difficulties in going from 'true' to 'truth', and Polanyi urges the importance of the distinction between *the probability of a statement* (corresponding to our 'It is probable that . . .') and *a probability statement* (a statement ascribing a probability, a statement *that* the probability

† See further below, Ch. XII.

of so-and-so is such-and-such).† But the difficulties in making meta-statements about probabilities are no greater than those in making meta-statements about truth; and in neither case give any support to subjectivism. Sophisticated probability-judgements, like unvarnished assertions of fact and sophisticated ascriptions of truth, although necessarily in the mouth of some particular speaker, are nevertheless trying to be objective, rational and true. Statements of probability do commit the speaker to something, though not to the same extent as simple statements of truth and falsehood. The concept of probability is a generalisation of the concepts of truth and falsehood. Probabilities come out of the same logical drawer as truth and falsehood, interpolated between them.

† Michael Polanyi, *Personal Knowledge*, London, 1958, Ch. 2, §§4–5, pp. 24–30.

III

THE MEASUREMENT OF PROBABILITIES

WE are seldom content with degrees. Once we can say that one thing is hotter than another, we start asking "How much hotter?". At the very least, where there are three things A, B, C, and B is hotter than C and less hot than A, we feel entitled to ask whether the difference between B and C is about the same as that between A and B, or much less or much larger. "Is B about half-way between A and C?" we ask; "or is it only a little hotter than C?", "or only a little less hot than A?" We ask these questions so insistently that we go to great artifices to answer them, often constructing purely conventional scales, as the Fahrenheit and Centigrade scales were when they were first introduced, in order to be able to give some sort of answer to these questions.

The same is true of probabilities. We want not only to compare probabilities but to assess them. We want to be able not only to say that it is more probable that the battle of the Halys was in 585 B.C. than that the battle of Marathon was in 490 B.C., and that this in turn is more probable than that the battle of Carchemish was in 606 B.C., but also to say that it is probable beyond all reasonable doubt that the battle of the Halys was in 585 B.C., and very probable that Marathon was in 490 B.C., but only moderately probable that Carchemish was in 606 B.C. These rough, non-numerical, estimates can often be asked for, and can sometimes be made. As we have already seen, they cannot always be asked for, because propositions can be of such widely differing sorts that they cannot be sensibly compared. Even when comparisons are possible, we may not be able to say which of two propositions is more probable; and even when we can, we may be unable to hazard any estimate of how much more probable the one is than the other. Nevertheless, we can sometimes. And, as we have already seen,† we have the beginnings of a scale in that there is a mid-point, when a proposition is as probable as not. The justification is a certain assumption of symmetry between truth and

† Ch. II, p. 11

falsehood. Except for the one difference, that true propositions
are to be believed and false ones disbelieved, truth and falsehood
enjoy parity of logical esteem. It is the same to say 'It is true that
q' or 'It is false that not q'.† And therefore if it is as probable that
q as that not q, the probability of q is as near truth as it is near
falsehood, and so half-way between.

We can go further, and sometimes do. But we are only estimat-
ing, and our estimates are inherently inexact. I can give it as my
opinion that it is ten to one that Marathon was in 490 B.C., and
only two to one that Carchemish was in 606 B.C., but if you think
that it is seven to three in favour of 606 B.C. for Carchemish, we
cannot sensibly try to resolve the difference. Our power to estimate
probabilities directly is very limited indeed. Let us therefore put
on one side, for the present, the problem of how we assign probabilities to propositions *etc.*, and assume that we can, and consider
what must follow. We shall return to the problem in Chapters V and
VIII and we shall find that in some cases we can assign them in-
directly, in virtue of what we shall already have shown to follow
from the various possible assignments. We therefore now turn to
the syntax, rather than the semantics, of probability-judgements.

If we are to be able to answer the question 'How much?', we
are committed to magnitudes with the order-type θ of the real
numbers.‡ In so far as probabilities are inherently inexact, it may
seem an idle disputation whether we need the rational numbers or
the real numbers to describe them. But in working out the syntax
of probability-judgements, we need to idealise: and in that case,
it is the real numbers we need.

Probabilities then are real numbers assigned to propositions,
propositional functions, statements, sentences, formulae, judgements, or anything else which could be intelligibly said to be true
or false. For the present we speak only of propositions, although
in Chapter IV we shall need to confine ourselves to propositional
functions. Propositions can be combined with other propositions
to form new, complex ones. Thus if we have the propositions q
and r, we can form their conjunction—or logical product—

$$q \ \& \ r \quad \text{(meaning both } q \text{ and } r)$$

† We avoid the conventional p, because it is sometimes used to refer not
to propositions but to probabilities.

‡ See E. V. Huntingdon, *The Continuum*, 2nd ed., New York, 1955,
Ch. V, esp. §54.

and their disjunction—or logical sum—

$$q \vee r \text{ (meaning either } q \text{ or } r \text{ or both).}$$

The whole propositional calculus can be developed from either one of these logical constants, together with negation, written $\sim q$ *etc.* (meaning not q).† The propositional calculus has a Boolean algebra.‡ If we are to have a calculus of probabilities which are magnitudes assigned to propositions, we have to devise some way of marrying the Boolean algebra of the propositions with the ordinary arithmetical algebra of the real numbers. The traditional calculus of probabilities is the offspring of that marriage; indeed, as I shall show it is essentially the only possible one.

Boolean algebra is like, but not exactly like, ordinary algebra. Readers not familiar with mathematical logic or elementary set-theory may find it helpful to use the following analogy: We consider the whole numbers, and divide them into two classes, odd and even. Odd will correspond to the value T in propositional calculus, I or V in set-theory: even to the value F in propositional calculus, 0 or $\emptyset$ or Λ in set theory. Adding one will correspond to negation in the propositional calculus, complementation in set-theory. Arithmetical multiplication will correspond to conjunction in propositional calculus, Boolean or "logical" multiplication in set-theory. For if we add one to an even number we get an odd number, and *vice versa*: and Even × Even is Even, Even × Odd is Even, Odd × Even is Even, and Odd × Odd is Odd, corresponding to the truth tables, $F \& F$ is $F, F \& T$ is F, $T \& F$ is $F, T \& T$ is T.§ The analogy reveals an important

---

† The propositional calculus can, indeed, be built up from *one* logical constant (*e.g.* neither q nor r); but this does not concern us here.

‡ Not only propositions but properties, features, propensities, qualities, and characteristics obey a Boolean algebra. The main argument of this chapter, although phrased in conformity with the thesis that probabilities are metalogical properties or values ascribed to propositions, is equally available if we view probabilities as propensities or any other sort of property. For the propensity theory, see K. R. Popper, "The Propensity Interpretation of Probability", *British Journal for the Philosophy of Science*, **10**, 1959, pp. 25–42; or D. H. Mellor, "Chance", *Proceedings of the Aristotelian Society*, Supplementary Volume, XLIII, 1969, pp. 11–36; or D. H. Mellor, *Chance*, Cambridge, forthcoming.

§ There is also a Boolean analogue of arithmetical addition (namely non-equivalence, or exclusive alternation, $q \not\equiv r$); but in view of the argument shortly to be given, it would be confusing to develop this analogy in detail.

difference between Boolean and ordinary arithmetical operations. With Boolean multiplication,

$$A \times A = A \quad \text{(or, in propositional calculus, } q \& q \equiv q)$$

and similarly, with Boolean addition,

$$A + A = A \quad \text{(or, in propositional calculus, } q \vee q \equiv q)$$

Boolean algebra has, so to speak, a built-in device for eliminating redundancy: but therefore its operations do not have corresponding to them any inverse operations of logical subtraction or logical division. These constitute important differences between Boolean and ordinary arithmetical algebras.

The traditional calculus of probabilities is based on three rules: one connecting probabilities with truth, the others connecting the probabilities of propositions with those of other propositions formed from them by means of the logical constants. The first rule assigns to true propositions the probability 1; we shall impose a slight qualification in Chapter V.† The second is the Negation Rule; it states:

If the probability of q is α, the probability of $\sim q$ is $1-\alpha$.

From these two rules together it follows that the probability to be assigned to false propositions is 0, and that all probabilities lie in the interval [0, 1].

For the third rule we have a choice, since the propositional calculus can be based either on conjunction—or logical multiplication—or disjunction—or logical addition—(or, indeed, on some other constants). If we base it on disjunction we need to give a rule to assign a probability to $q \vee r$ (either q or r or both) in terms of the probabilities assigned to q and to r. But this proves impossible in general, because of the 'or both' implicit in the disjunction. We do not want to double-count probabilities, but the definition of disjunction allows for there being an indefinite degree of double-reckoning. We may exclude this by *fiat*. We can lay down that the rule for disjunction shall apply only where q and r exclude each other, so that there is in fact no possibility of both q and r, and $q \vee r$ can, in these cases, be read as either q or r full

† pp. 85–94.

stop. With such a proviso, we can formulate a restricted Disjunction Rule; it states:

if the probability of q is α, and of r is β, and provided q and r are mutually exclusive, then the probability of $q \vee r$ is $\alpha + \beta$.

or, more mnemonically,

the probability of the logical sum of two mutually exclusive propositions is the arithmetical sum of their probabilities.

It should be noted that we can almost obtain the Negation Rule from the Disjunction Rule; for q and $\sim q$ are logically exclusive of each other, and the probability of $q \vee \sim q$ is 1 by Rule I. But, for all we have said so far, probabilities might be negative, or greater than 1, and these would fail to satisfy our requirement that probabilities lie *between* truth and falsehood. It is possible to stipulate that probabilities should not be negative. This is what is done by Cramér, who gives the following three axioms†:

   I. Any probability $P(A)$ is a non-negative number: $P(A) \geqslant 0$.
   II. The probability of a certain event is equal to unity.
   III. If the events A and B are mutually exclusive, we have the addition rule: $P(A + B) = P(A) + P(B)$.

Kolmogorov gives a more formal axiomatization, beginning with the following five‡:

Let E be a collection of elements $\xi, \eta, \zeta, \ldots$, which we shall call *elementary events*, and $\mathscr{F}$ a set of subsets of E; the elements of the set $\mathscr{F}$ will be called *random events*.

   I. $\mathscr{F}$ is a field of sets.
   II. $\mathscr{F}$ contains the set E.
   III. To each set A in $\mathscr{F}$ is assigned a non-negative real number $P(A)$. This number $P(A)$ is called the probability of the event A.
   IV. $P(E)$ equals 1.
   V. If A and B have no element in common, then $P(A + B)$ $= P(A) + P(B)$.

The first two axioms lay down the Boolean algebra of "random events", with E being the "universal event". Axiom III states

† H. Cramér, *The Elements of Probability Theory*, New York and Stockholm, 1955, p. 33.

‡ A. N. Kolmogorov, *Foundations of the Theory of Probability*, tr. N. Morrison, New York, 1956, Ch. I, §1, p. 2.

that probabilities are real numbers, *and* non-negative. Axiom IV corresponds to our Rule I; axiom V to our Rule III.

There are many other axiomatizations of probability: but Kolmogorov's may be regarded as the paradigm. The probability calculus is a part of set-theory—additive set-theory; axiom III assigns non-negative real numbers to sets, and axiom V enables them to be added together. From these axioms, together with a definition of "conditional probability"† and an axiom of continuity needed for infinite systems of sets,‡ a rigorous and wide-ranging theory can be developed. But, as with all formal theories, the questions remain: What has this theory to do with probability? Why should we accept the axioms? or at least, Why should we take the axioms as giving an implicit definition of probability? The axiomatic approach, for all its rigour and elegance, is no use unless it axiomatizes the right concept. Otherwise, it remains simply a branch of pure mathematics, very interesting to those who are interested in additive set-theory, but of no relevance to those concerned with partial truth. Probability is a gerundive concept. It indicates how much propositions are to be believed. We cannot redefine it as that which is defined by Kolmogorov's axioms, any more than we can redefine good as that which is desired or is conducive to the greatest happiness of the greatest number, without thereby committing a naturalistic fallacy. If we are to use either Kolmogorov's or any other axioms for the probability calculus, we must show that they are appropriate for the calculation of partial truth.

In fact, we can justify the axioms of probability theory, granted certain assumptions. The earliest treatments of probability theory in the Seventeenth and Eighteenth Centuries started from an assumption of equiprobable case-types, from which all three of our Rules were proved. The Frequency theory starts from a certain sort of large classes ("Collectives") of case-instances, of which a certain fraction (or proportion or frequency) have the property we are interested in. Professor Braithwaite has offered a theory which combines some features of both.§ The mathematical structure is the same in each case. It is the theory of proper fractions. Once we have defined probability as the ratio of

---

† *Op. cit.*, Ch. I, §4, p. 6.   ‡ *Op. cit.*, Ch. II, §1, p. 14.
§ R. B. Braithwaite, *Scientific Explanation*, Cambridge, 1953, Chs. V, VI, and VII.

favourable cases to all cases, or the frequency with which cases with the property in question occur, we can apply the theory of proper fractions in a simple and intuitive way to yield the normal rules of the probability calculus. But the initial definitions themselves are open to objection. With the Equiprobability theories, it is the assumption of equiprobability between case-types that is in need of justification. Such a justification is not, as some writers aver, always unobtainable. But it is not always obtainable, and therefore does not provide a satisfactory basis for a definition of probability. The Frequency theory makes no *a priori* assumptions of probability. It attempts to deal entirely *a posteriori* with observed frequencies of event-instances. But since there are indefinitely many of these, there is difficulty in applying the theory of fractions at all, for fractions have definitely finite numerators and denominators; we can work out a perfectly satisfactory class-ratio arithmetic for any given finite classes, but we cannot easily, or perhaps at all, extrapolate it to indefinitely large classes, and therefore the Frequency theory too does not provide a non-problematic buttress for the axiomatic approach. We shall need to consider both theories more fully later on.† For the present, it is enough that there is no easy supplement available from class-ratio arithmetic which would render the axiomatic approach adequate in its own right.

We can give a less ambitious justification of the axioms, which does not depend on our defining probability in either Equiprobability or Frequency terms, but is inherent in any concept of probability as the idealisation of a guarded guide into a scale of partial truth. For then the probability calculus must be able to deal not only with probabilities lying in between truth and falsehood, but with the extreme cases as well; and not many rules will do this satisfactorily. Any rule of negation must, moreover, satisfy the Law of Double Negation, that $\sim \sim q \equiv q$. Hence any arithmetical operation R that is to be the analogue of negation must be such that $R^2 = I$, where I is the identity operator and hence $R = R^{-1}$. Multiplying by -1 is one such operation, taking the reciprocal is another. But the latter runs into difficulties with zero divisors. Hence we do well to build round the former: and our Negation Rule, $(1 - \alpha)$, is one based on this

† The Frequency theory in Ch. V, pp. 95–100; the Equiprobability theory in Ch. VII.

principle. We can argue with a fair show of reason for replacing the F of the propositional calculus by 0, since in set-theory the analogue of F is the "null set" or "empty set". We then obtain 1 as the probability of true proportions. We cannot obtain an immediate analogue of the disjunction rule, because logical addition differs, in having $1 + 1 = 1$, from ordinary arithmetical addition where $1 + 1 = 2$. But we can have a natural conjunction rule, since $1 \times 1 = 1$, and $0 \times 1 = 1 \times 0 = 0 \times 0$, according to arithmetic rules, which is just what we want for our Boolean purposes. Hence we might hope to provide some sort of justification for a calculus of probabilities assigning 0 to false propositions, and using the Negation Rule and the Conjunction Rule, which states:

If the probability of q is α, and of r is β, and provided q and r are independent, then the probability of q & r is $\alpha \times \beta$.

or, more mnemonically,

The probability of the logical product of two independent propositions is the arithmetical product of their probabilities.

The proviso is important. With the Disjunction Rule we had to lay down a subsidiary application rule that it was to be applied only when the propositions q and r were exclusive. With the Conjunction Rule we need a comparable application rule that we apply it only when the propositions q and r are *independent*;† for, obviously, if whenever q was true r was true also, the probability of q *and* r would be no different from that of q alone. We give no further definition of independence. Normally, indeed, we make the rule work backwards, and say that q and r *are* independent if and only if the Conjunction Rule is applicable. It is noteworthy that the independence rule for Conjunction is markedly different from the exclusiveness rule for Disjunction. One advantage of the Conjunction-plus-Independence approach over that of Disjunction-plus-Exclusiveness is that later we shall be able to drop the requirement of independence,‡ whereas dropping the requirement of exclusiveness makes the Disjunction Rule more complicated.

The intuitive justification does not go very far. The whole point

† In some books the terms 'statistically independent' or 'stochastically independent' are preferred.

‡ Ch. IV.

of probability theory is that it applies between the extremes of truth and falsehood, and all that we have established is that traditional probability theory satisfies one necessary condition of being correct, but not a sufficient condition. Moreover, it is easy to construct artificial variants of the traditional calculus which satisfy the condition required just as well. We could make false propositions have the value 1 and true propositions the value 2, and the Negation Rule:

If the probability of q is α, the probability of $\sim q$ is $3 - \alpha$,

and the Conjunction Rule:

If the probability of q is α, and of r is β, and provided q and r are independent, then the probability of $q \& r$ is $\alpha\beta - \alpha - \beta + 2$.

It is easy to produce indefinitely many variants. But there is a certain sameness about them. They are all obtained from our original three rules by some algebraic transformation. We might say that, in spite of their different forms, they were all essentially the same, all variants on the one canonical form that we originally gave.

The question then becomes whether there is any other family of rules which will satisfy the necessary conditions for applying to probabilities, or whether the family based on the three rules we originally gave is the only one which could give a satisfactory calculus of probabilities. I shall show that the latter alternative holds, and that once we have granted that probabilities are continuous degrees between truth and falsehood, and that the Boolean algebra of propositions must be married to the arithmetic algebra of magnitudes, there is essentially only one set of rules which will satisfy the conditions we shall be obliged to stipulate.

We deal with the Conjunction Rule first. There are four different arguments we can use. The shortest is a somewhat abstract one borrowed from the theory of groups. Probabilities under the operation of conjunction form a "semi-group with unit element" or "monoid". For

(i) If q has the probability $\mathrm{pr}(q)$† and r has the probability $\mathrm{pr}(r)$, then $q \& r$ is uniquely defined, and has some unique probability $\mathrm{pr}(q \& r)$ assigned to it.

† We write $\mathrm{pr}(q)$ for the real number assigned at the outset as a probability-value to the proposition q, *etc.* We call these assignments "preliminary probabilities", in contrast to the ones we shall assign after regraduation, which we shall, on p. 43, symbolize by $\mathrm{Prob}[q]$ *etc.*

(ii) The associative law holds for the operation of conjunction, for $\mathrm{pr}(q \mathbin{\&} (r \mathbin{\&} s))$ must be the same as $\mathrm{pr}((q \mathbin{\&} r) \mathbin{\&} s)$.

(iii) The probability of a true proposition must act as a unit element under the operation of conjunction, for if q is true and r is any other proposition,

$$\mathrm{pr}(q \mathbin{\&} r) = \mathrm{pr}(r) \quad \text{and} \quad \mathrm{pr}(r \mathbin{\&} q) = \mathrm{pr}(r).$$

Any operation and set of entities satisfying these three conditions constitutes a semi-group with unit element. We have already seen that there is no inverse operation to conjunction, and therefore probabilities under conjunction do not form a group.† They do, however, satisfy two further conditions. We have, first, required that they satisfy certain requirements of continuity, not because we could ever conceive ourselves making absolutely accurate assessments of probability, but because continuity is implicit in our ideal of probabilities being magnitudes.‡ And secondly, not only truth but falsehood plays a peculiar *rôle*, and probabilities have what I might term a "universal element", namely the probability of a false proposition, for

(iv) if q is false and r is any other proposition, then

$$\mathrm{pr}(q \mathbin{\&} r) = \mathrm{pr}(q) \quad \text{and} \quad \mathrm{pr}(r \mathbin{\&} q) = \mathrm{pr}(q).$$

In view of conditions (i), (ii), (iii) and considerations of continuity, we can consider probabilities under conjunction as a particularly simple type of semi-group, a continuous semi-group with one parameter and with a unit element. Now there is a general theorem about continuous groups, applicable also to semi-groups with a unit element, which states that any one-parameter group of transformations is equivalent to a one-parameter group of translations.§ The group operator is thus equivalent to $+$ and the unit element to 0. The universal element can then be identified as being equivalent to $\pm\infty$, and in our special case of a semi-group we can identify it as $-\infty$. This semi-group can be transformed again by exponentiation and shown to be equivalent to the group of real numbers from 0 to 1 under the operation of multiplication, with 1 as the unit element and 0 as the universal element.

† p. 26 above.   ‡ See above, p. 24, and below, p. 34.
§ See, *e.g.*, Luther P. Eisenhart, *Continuous Groups of Transformations*, Princeton, 1933, p. 34.

The equivalence theorem for one-parameter groups of transformations is of interest because it has other applications in our philosophy of nature, and can be used to show that assumptions of additivity or uniqueness which are often made are in fact justifiable.† To prove it and then apply it to the Conjunction Rule would, however, be using a steam-hammer to crack a nut. Not only is the proof fairly long, but most of it is not needed, since probabilities are very much simpler than the many-dimensional manifolds to which the theorem can apply. Instead, I give a direct proof that if we have any plausible rule for calculating the probabilities of a conjunction of propositions, then we can produce a "re-graduating function", which will re-graduate our original or "preliminary" probabilities in such a way that the original rule of conjunction, whatever it was, is transformed for the regraduated probabilities into the canonical Rule of Conjunction that we have given. All possible rules of conjunction are thus isomorphic with our Rule of Conjunction, which, of them all, is expressed in the simplest and most perspicuous, and hence canonical, form.

No assignment of magnitudes to propositions and conjunction rule would be plausible unless it satisfied certain conditions, some of them arithmetical and some Boolean. Probabilities lie in between truth and falsehood: no assignment would be plausible if it assigned to any proposition a magnitude that did not lie between that assigned to true propositions and that assigned to false propositions. Therefore whatever values we assign to true and false propositions, we must not assign to any proposition a value outside these limits. We may, if we like, set the extremes at $-\infty$ and $+\infty$, in which case every real number will be a permissible value for a probability; but unless we do this, we must restrict our assignment of probabilities to keep them within the interval bounded by the values assigned to true and to false propositions. It is convenient also to introduce the convention that the magnitude assigned to true propositions is greater than that assigned to false propositions. It is only a convention. All that is non-conventional is that the magnitude assigned to true propositions is different from that assigned to false, so that it is either greater or

† I use this theorem in *A Treatise on Time and Space*, (forthcoming) to argue that natural laws can be expressed in a form that is independent of time.

less. But it makes the argument simpler if we say that it is greater; and at any stage we could change the convention by multiplying by −1. We therefore lay down as our first condition

(i) There are no probabilities greater than that of a true proposition or less than that of a false proposition.

The Conjunction function, whatever it is, must be a function of two real numbers—the probabilities of the conjoined propositions —and for all pairs of values that are not outside the limits set by condition (i), must have one and only one value, also not outside the limits set by condition (i). For if every proposition is to have a probability, there must always be a probability assigned to a proposition which is a conjunction of propositions; and this, like all other assignments, must be unique.

The Conjunction function must be continuous. If we have two conjunctions, $q\,\&\,s$ and $r\,\&\,s$, with s the same in both, then if the probability assigned to q were very close to that assigned to r, we should expect the probability assigned to the conjunction of q with s to be very close to that assigned to the conjunction of r with s. In practice, as we have noted, we hesitate to assign probabilities with any high degree of accuracy, and would be reluctant to say whether the probability of q was very close to that of r, or whether the probability of $q\,\&\,s$ was very close to that of $r\,\&\,s$. But when we idealise our intuitive notion of probabilities by assigning real numbers to propositions, we can no longer avoid the question. If a Conjunction function were proposed which was discontinuous at some point, it would be implausible. And therefore we lay down a condition of continuity for conjunction:

(ii) As the probability assigned to q approaches that assigned to r, so must the probability assigned to $q\,\&\,s$ approach that assigned to $r\,\&\,s$.

The next condition is a condition of what the mathematicians call monotonicity, and is intuitively more acceptable. Whereas continuity is concerned with assignments of probability being almost the same, monotonicity is concerned with their being different. It states that the value of the probability of a conjunction must differ in the same direction, though not necessarily to the same extent, as the value of the probability of the conjoined propositions. That is, the probability of a conjunction increases if the probability of either of the conjoined propositions increases; or in symbols,

(iii) If $\mathrm{pr}(q) > \mathrm{pr}(r)$, $\mathrm{pr}(q\,\&\,s) > \mathrm{pr}(r\,\&\,s)$, except when s is false.

The Boolean conditions are those already specified. We have first the requirement of associativity which we can state as

(iv) $\mathrm{pr}(q \,\&\, (r\,\&\,s)) = \mathrm{pr}((q\,\&\,r)\,\&\,s)$.

We have next the special *rôle* of true propositions, that the conjunction of a true proposition with any other should have the same probability as that other proposition; or in symbols

(v) if q is true, $\mathrm{pr}(q\,\&\,r) = \mathrm{pr}(r)$.

Commutativity is the most intuitively obvious requirement for any conjunction rule. If a conjunction rule were suggested which assigned a different probability to $q \,\&\, r$ from that assigned to $r\,\&\,q$, we should have no hesitation in rejecting it out of hand. The conjunction rule must be symmetrical in its two arguments. I find it somewhat surprising that this condition is not required in the mathematical argument I am about to give; indeed, if we put the Associative Law in another commonly used form

(iv') $\mathrm{pr}(q\,\&\,(r\,\&\,s)) = \mathrm{pr}(r\,\&\,(q\,\&\,s))$,

the Commutative Law obviously follows from that together with (v); and, in fact, it must hold even if we only have (iv), together with (i)–(iii) and (v).

The special *rôle* of false propositions is the final Boolean condition.

(vi) If q is false, $\mathrm{pr}(q\,\&\,r) = \mathrm{pr}(q)$.

It is because of this we need the proviso in condition (iii). We should note the difference between (v) and (vi). Conjunction with a true proposition leaves the probability the same as the other proposition: conjunction with a false proposition makes the probability the same as the false one. It may be thought strange that true and false propositions have different *rôles*, since we have already argued,† and shall argue more stringently,‡ that they are symmetric. The answer is that they are symmetric, but not with respect to conjunction. If we apply the principle of duality to Boolean algebra, we have to interchange not only true and false, but conjunction and disjunction as well. While we are dealing with possible conjunction rules, true and false propositions are significantly different: conjunction with a true proposition is an identity operation, and we can regard the probability of true propositions as a unit element: conjunction with a false proposition has universally the same result, and we can regard the probability of a false proposition as a "universal" element. It is because there are both, that all conjunction rules must be essentially the same.

There are several different ways in which the proof can be accomplished, each starting from slightly different assumptions.§ All require Associativity (condition (iv)) and Continuity (condition (ii)). The simplest proof, which I shall give, ¶ requires a slightly stronger condition than that the conjunction function should be continuous, namely that it should be differentiable, with continuous first-order derivatives with respect to both‖ variables; it also requires Monotonicity (condition (iii)), and the special properties of true propositions (condition (iv)). Alternatively, we can dispense with the special properties of true propositions if we have Commutativity as well as Associativity, and keep the same conditions of Monotonicity and Continuity-plus-

---

† pp. 23-4.   ‡ p. 44.

§ The non-mathematical reader can skip the mathematical argument, and taking on trust the theorem finally proved on p. 41, resume reading on p. 43.

¶ I owe this proof to my former colleague, Dr. M. S. P. Eastham. It is an adaption of arguments originally due to G. J. Whitrow, *Quarterly Journal of Mathematics*, 6, 1935, para. 4, pp. 252–6; E. A. Milne and G. J. Whitrow, *Zeitschrift für Astrophysik*, 15, 1938, pp. 273–4; A. G. Walker, *Quarterly Journal of Mathematics*, 17, 1946, pp. 67–8; G. J. Whitrow, *The Natural Philosophy of Time*, Edinburgh, 1961, p. 172. W. H. M'Crea, *Proceedings of the Royal Irish Academy*, 45, 1938, A, pp. 24–5.

‖ If we have already established that the conjunction function is commutative, we do not need to stipulate both. But unless we have stipulated Associativity in the stronger form (iv′) instead of (iv), we shall not have established Commutativity until we have proved the first stage of the theorem.

Differentiability.† Alternatively, again, Aczél‡ needs only Continuity in the minimal sense and Monotonicity, and does not need to postulate a unit operator so long as he has an analogue of our condition (i), namely that the Conjunction function is always defined within a certain interval and always has its value in that interval. Cox§ does not need condition (i), nor condition (iii) (Monotonicity), and only requires (v) (the special properties of true propositions) to identify the unit element after the Conjunction function has been reduced to simple addition. But he needs an even stronger requirement of continuity, namely that the Conjunction function should not only be continuous and differentiable, but have *second*-order derivatives that are continuous too. It is still a reasonable stipulation. It would be unreasonable to appeal to our intuitions to justify it, but it is fair to observe that it is a condition of respectability for mathematical functions; and while there are functions which, although continuous, are not differentiable, or have discontinuous derivatives, they are not often applicable to our understanding of the world, and that unless there was some special reason for using one of them, it would be unreasonable to do so.

We can reformulate and condense the conditions required for our proof. We first define pr(q) as a one-valued function of propositions having its value a real number in some closed interval [v, w] where $-\infty \leqslant v < w \leqslant \infty$, and w is the value assigned to all true propositions and v the value assigned to all false propositions. We then define Con $\{x, y\}$ as a single-valued function defined for all pairs of real numbers within the closed interval [v ,w] and having its values within that interval, satisfying the following four conditions:

(i) Con $\{x,$ Con $\{y, z\}\}$ = Con $\{$Con $\{x, y\}, z\}$

(ii) Con $\{w, x\}$ = x for all x

(iii) Con $\{v, x\}$ = v for all x

---

†These were the conditions required by Abel, who was the first to devise a regraduating function. I am indebted to Dr. I. J. Good, of Trinity College, Oxford, for bringing this to my notice. Good himself uses a regraduating function in Appendix 3 of his *Probability and the Weighing of Evidence*, London, 1950.

‡ Jean Aczél, "Sur les Opérations Definies pour Nombres Réels", *Bull. Soc. Math. Française*, **76**, 1948, pp. 59–64.

§ See note on p. 41 for references.

(iv) $\dfrac{\mathrm{d}\,\mathrm{Con}\,\{x,\,y\}}{\mathrm{d}x}$ and $\dfrac{\mathrm{d}\,\mathrm{Con}\,\{x,\,y\}}{\mathrm{d}y}$ exist, are positive provided $y \neq v$ or $x \neq v$ respectively, and are themselves continuous in x and y.

Condition (iv) as now stated includes the requirement of Monotonicity as well as those of Continuity.

It is easy to see that whatever the Conjunction function may be, if it is differentiable its derivations will have to have the

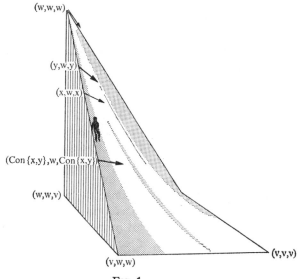

Fig. 1.

canonical values at the extremes of truth and falsehood; that is, since

$$\mathrm{Con}\,\{w,\,x\} = x \quad \text{and} \quad \mathrm{Con}\,\{v,\,x\} = v$$

$$\frac{\mathrm{d}\,\mathrm{Con}\,\{w,\,x\}}{\mathrm{d}x} = 1 \qquad \frac{\mathrm{d}\,\mathrm{Con}\,\{v,\,x\}}{\mathrm{d}x} = 0.$$

This suggests that gradients may provide a key, and indeed they do. We discover the regraduating function by considering the way in which probabilities of conjunctions must behave when one of the conjoining propositions approaches the value True. The non-mathematician may find it helpful to imagine himself at the "summit" of the conjunction function, which, by Monotonicity, must be where both the conjoined propositions have the value True. He is then to descend down the line along which one of the conjoined propositions retains the value True while the other is assigned lower and lower probabilities. The values of the conjunction function must also diminish, but they will form a sort of ridge in that for any assigned probability of the other function, the value of the conjunction is at a maximum when the first proposition has the value True. We shall consider how, as one goes *along* the ridge, the gradient in the other direction going down *away* from the ridge, varies; in particular we shall consider the ratio of the gradient at any point along the ridge to the parallel gradient at the very summit. This ratio will yield the required regraduating function.

It is the Associative Law which gives the point of entry into the argument. For if a conjunction is itself conjoined with a true proposition, it is the same as if one of the conjoined propositions had been conjoined with the conjunction of the other with the true one. The reader should not attempt to visualise these functions of three variables, but follow the formal argument, that since, by condition (i),

$$\text{Con}\,\{x,\,\text{Con}\,\{y,\,z\}\} = \text{Con}\,\{\text{Con}\,\{x,\,y\},\,z\}, \tag{1}$$

if, for any fixed x, we differentiate with respect to z, we shall have

$$\frac{\mathrm{d}\,\text{Con}\,\{x,\,\text{Con}\,\{y,\,z\}\}}{\mathrm{d}\,\text{Con}\,\{y,\,z\}} \cdot \frac{\mathrm{d}\,\text{Con}\,\{y,\,z\}}{\mathrm{d}z} = \frac{\mathrm{d}\,\text{Con}\,\{\text{Con}\,\{x,\,y\},\,z\}}{\mathrm{d}z}. \tag{2}$$

When $z = w$, the second factor of the left-hand side gives the gradient of $\text{Con}\,\{y,\,z\}$ at the ridge; it is a function only of y, and can be written, for brevity's sake, $\Omega(y)$. The right-hand side similarly gives the gradient of $\text{Con}\,\{\text{Con}\,\{x,\,y\},\,z\}$ at the ridge, and can be written $\Omega(\text{Con}\,\{x,\,y\})$. The first factor on the left-hand side can also be simplified, since y appears in it only inside the function $\text{Con}\,\{y,\,z\}$, which is itself equal to y when $z = w$. The first factor therefore becomes

$$\frac{\mathrm{d}\,\text{Con}\,\{x,\,y\}}{\mathrm{d}y}$$

and the whole equation can be written

$$\frac{\mathrm{d}\,\text{Con}\,\{x,\,y\}}{\mathrm{d}y} \cdot \Omega(y) = \Omega(\text{Con}\,\{x,\,y\}) \tag{3}$$

which can be rewritten

$$\frac{d \operatorname{Con} \{x, y\}}{\Omega(\operatorname{Con} \{x, y\})} = \frac{dy}{\Omega(y)} \tag{4}$$

since Ω is positive in $(v, w]$ by condition (iv).

Equation (4) tells us something about how the gradient away from the ridge alters as one goes along the ridge, in view of the Associative Law. For any given x, the two points on the ridge, y and $\operatorname{Con} \{x, y\}$ (which, for a given fixed x, will depend on y alone), must satisfy the differential equation (4). They are represented, for a particular y, by the bold arrows on figure 1. We then integrate (4), taking the definite integral along the ridge down from the summit $y = w$, and obtain

$$[\log \Omega(\operatorname{Con} \{x, y\}) - \log \Omega(\operatorname{Con} \{x, w\})]$$
$$= [\log \Omega(y) - \log \Omega(w)]. \tag{5}$$

Each side of equation (5) expresses the logarithm of a ratio of gradients at different points along the ridge: on the right-hand side the ratio of the gradient at y to the gradient at w, on the left-hand side the ratio of the gradients at the corresponding points under conjunction with x, namely $\operatorname{Con} \{x, y\}$ and $\operatorname{Con} \{x, w\}$. But $\operatorname{Con} \{x, w\} = x$, by (ii). Moreover, at the summit,

$$\Omega(w) = \left[\frac{d \operatorname{Con} \{w, z\}}{dz}\right]_{z=w} = \left[\frac{dz}{dz}\right]_{z=w} = 1$$

and

$$\log \Omega(w) = 0.$$

Hence we can simplify (5) to

$$\log \Omega(\operatorname{Con} \{x, y\}) = \log \Omega(x) + \log \Omega(y) \tag{6}$$

and so, taking exponentials,

$$\Omega(\operatorname{Con} \{x, y\}) = \Omega(x) . \Omega(y). \tag{7}$$

We already have proved that $\Omega(w) = 1$; we now consider condition (iii),

$$\operatorname{Con} \{v, x\} = v \quad \text{for all } x.$$

Hence, differentiating with respect to x:

$$\frac{d \operatorname{Con} \{v, x\}}{dx} = 0$$

and so

$$\left[\frac{d \operatorname{Con} \{v, x\}}{dx}\right]_{x=w} = 0;$$

that is,

$$\Omega(v) = 0.$$

Thus

$$\Omega(x), \quad i.e. \quad \left[\frac{d \, \text{Con} \, \{x, z\}}{dz}\right]_{z=w},$$

is the regraduating function we require; and, whatever our preliminary assignment of probabilities, we can regraduate them so
as to have the Conjunction Rule become a simple product rule,
with 1 as the probability of true propositions and 0 as the probability of false ones.

Cox's method is much the same.† He, too, finds an $\Omega(y)$ which
satisfies equation (3):

$$\Omega(\text{Con} \, \{x, y\}) = \frac{d \, \text{Con} \, \{x, y\}}{dy} \cdot \Omega(y),$$

but he finds it by working out the partial *second*-order derivatives
that can be obtained from the Associative Law, and eliminating
all functions of Con $\{x, y\}$ and Con $\{y, z\}$ other than their derivatives. This yields an equation both of whose sides can be integrated with respect to y, and which can be expressed in the form

$$\frac{\partial}{\partial y} \log \left[\frac{\dfrac{\partial \, \text{Con} \, \{x, y\}}{\partial x}}{\dfrac{\partial \, \text{Con} \, \{x, y\}}{\partial y}}\right] = - \frac{\partial}{\partial y} \log \left[\frac{\dfrac{\partial \, \text{Con} \, \{y, z\}}{\partial y}}{\dfrac{\partial \, \text{Con} \, \{y, z\}}{\partial z}}\right]. \quad (8)$$

The crux of Cox's argument is that in this equation x appears
only on the left-hand side, z only on the right-hand side, and
therefore both left- and right-hand sides must in fact be functions
only of y. We can therefore define

$$\Omega(y) = \frac{\dfrac{\partial \, \text{Con} \, \{x, y\}}{\partial x}}{\dfrac{\partial \, \text{Con} \, \{x, y\}}{\partial y}} \cdot C$$

where C is a constant of integration, and from this, making use of
the right-hand side of (8), Cox is able to derive equation (3).

† R. T. Cox, "Probability, Frequency and Reasonable Expectation",
American Journal of Physics, Vol. XIV, 1946, pp. 6, 11–12; see also *The
Algebra of Probable Inference*, Baltimore, 1961, Ch. I, pp. 14ff., for a
slightly different handling of the differentials.

The essential difference between the two approaches is that Cox considers the ratio between the gradients of Con $\{x, y\}$ in both directions anywhere, whereas I consider only the one gradient as we reach the "ridge" of maximum values (for each fixed y) of Con $\{y, z\}$ which are attained as z approaches w; or, to be exact, the ratio of the gradient anywhere along the ridge to the gradient at the ridge's own summit, when $y = w$ and z approaches w, as compared with the ratio of the gradients at the corresponding points along the ridge obtained by conjoining each of the original points y and w with some third, say x. The assumption that there is such a ridge saves a lot of heavy algebra and fixes various constants of integration. But essentially both proofs turn on having the Associative Law and enough differentiability, from which we can extract relations between the complete differential d Con $\{x, y\}$ and dx and dy. The integration involves natural logarithms, which lead us immediately to a multiplication, or product, rule.

Both these arguments depend on the Conjunction function being differentiable, and we may still wonder whether we are entitled to make any such assumption. It is a merit of Aczél's proof that he does not need to make it. He shows that if there is an operation on two variables defined in the interval (a, b) and having its result always in that interval, and satisfying the three conditions:

   I. Monotonicity: $x \circ y < x' \circ y$ for $x < x'$ (and the same for $y < y'$);

  II. Continuity: $\lim (x \circ y) = (\lim x) \circ (\lim y)$;

 III. Associativity: $(x \circ y) \circ z = x \circ (y \circ z) = x \circ y \circ z$;

then there is a strictly increasing, continuous function $f(z)$ defined in the interval (a, b) such that

$$x \circ y = f^{-1}[f(x) + f(y)].$$

Aczél constructs his $f(z)$ by iterating the operation $\circ$ with the same argument. He defines $F_n(x)$ as $x \circ x \circ x \circ \cdots \circ x$ n times, and $F_{1/n}(x)$ as its inverse. He thus can define $F_{p/q}(x)$ for all rational numbers p/q, and defines $\phi(p/q)$, the inverse of the regraduating function, for each rational number p/q as $F_{p/q}(A)$ for a certain fixed A. In view of Monotonicity he is able to show that there can be only one unit element, and to extract useful consequences about the effect of the operator as one of its arguments approaches the upper bound, b, of the interval,† which show that $\phi_{p/q}$ will have the properties required. Having been defined for all rational numbers, $\phi(x)$ can by Continuity be extended to all real numbers. The inverse of $\phi(x)$ maps the original interval into the real numbers with 0 as the unit element, and can be shown to transform the operation ∘ into simple addition $+$. This gives us the essential step, and it is easy then to identify the unit and universal elements and thence by exponentiation arrive at the canonical form of the Conjunction Rule.

Any of these proofs is sufficient to establish the theorem: If the four conditions on pp. 37–8 hold, there is a regraduating function, which I shall write $f(x)$ for short, which is such that

$$f(\text{Con } \{\text{pr}(q), \text{pr}(r)\}) = f(\text{pr}(q)) \times f(\text{pr}(r)),$$

and $f(w) = 1$,

and $f(v) = 0$.

Hence, if we rewrite our regraduated preliminary probabilities, f $(\text{pr}(q))$, as $\text{Prob}[q]$ etc., we shall have

$$
\begin{aligned}
\text{Prob}[q \, \& \, r] &= f(\text{pr}(q \, \& \, r)), \\
&= f(\text{Con } \{\text{pr}(q), \text{pr}(r)\}) \\
&= f(\text{pr}(w)) \times f(\text{pr}(r)) \\
&= \text{Prob}[q] \times \text{Prob}[r];
\end{aligned}
$$

which is our original Conjunction Rule, that the probability of the logical product of two independent propositions is the arithmetical product of their probabilities.

Once we have established the canonical form of the Conjunction Rule, the Negation Rule will have to satisfy the following conditions:

(i) The probability of the negation of any proposition must lie in the interval [0, 1].

(ii) The probability of the negation of the negation of any proposition is the same as that of the proposition itself.

(iii) The probability of false propositions, which are the negation of true propositions, is 0.

† Aczél actually considers the opposite case—the lower bound, supposing there is any A such that $A \circ A > A$. But in probability we are going to assume $A \circ A < A$, and it is on the frontier with truth that probability acquires its special properties.

These three conditions suggest that the Negation Rule must be:

If the probability of q is α, the probability of $\sim q$ is $1 - \alpha$.

For if we have a Negation function, $\text{Neg}(x)$, which yields the probability of $\sim q$ given the probability of q, condition (ii) above rules out all functions except those based on multiplication by -1 or on reciprocals; condition (i) rules out the latter, and condition (iii) specifies it to the form $1 - x$. A rigorous proof can be given, granted these three conditions, together with one other intuitively acceptable one, and a fifth, more questionable one. Let us restate them:

(i) $\text{Neg}(x)$ is continuous and differentiable in the interval $[0, 1]$;

(ii) $\text{Neg}[\text{Neg}(x)] = x$;

(iii) $\text{Neg}(0) = 1$;

(iv) $\dfrac{d}{dx} \text{Neg}(x)$ is never positive;

(v) $\text{Neg}'(x) = \text{Neg}'[\text{Neg}(x)]$, where $\text{Neg}'(x) = \dfrac{d}{dx} \text{Neg}(x)$.

The last condition represents a certain duality truth and falsehood, which is embodied in the principle of duality in the propositional calculus. Here we are extending this principle to the metric of any measure of probability, and saying that the rate of increment of the probability of any proposition and the rate of decrease of the probability of its negation must be equal: $d\,\text{Neg}(x)/dx$ is the same for $\text{Prob}[\sim q]$ as for $\text{Prob}[q]$. This seems to me to be a metrical analogue of the principle that so far as negation goes, there is a complete isomorphism between the calculus of propositions and its transform under the operation of negation, and it is only an arbitrary assignment that picks out one truth value as true and the other as false. So too with probability, apart from the "directional" correlation of increase of probability-value with greater degree of truth, there is nothing else to distinguish the measure of probability at $\text{Prob}[q]$ from that at $\text{Prob}[\sim q]$. However, I put this forward only rather tentatively.†

† Cox is able to show without recourse to condition (v) that, provided we already have a rule of conditional probability, $\text{Neg}(x)$ is of the form $(1 - x^m)^{\frac{1}{m}}$ where m is an arbitrary constant. But I have not been able to show that m must have the value 1 without some further assumption. See below, Ch. IV, pp. 67–9.

If condition (v) be granted, then it will follow, very easily, that $\text{Neg}(x) = 1 - x$. For, differentiating (ii), which we can do by (i), we have

$$\text{Neg}'(\text{Neg}(x)) \times \text{Neg}'(x) = 1.$$

Hence, from (v)

$$\text{Neg}'(x) \times \text{Neg}'(x) = 1$$
$$\therefore \text{Neg}'(x) = \pm 1.$$

Therefore by (iv)

$$\text{Neg}'(x) = -1$$
$$\therefore \text{Neg}(x) = -x + C$$

where C is some constant;
but by (iii)

$$\text{Neg}(0) = -0 + C = 1$$
$$\therefore C = 1$$
$$\text{Neg}(x) = 1 - x.$$

To obtain the Disjunction Rule for independent propositions we use one of De Morgan's Laws, which states that

$$q \vee r \quad \text{is equivalent to} \quad \sim(\sim q \,\&\, \sim r).$$

If the probability of q is α, and of r is β, then the probabilities of $\sim q$ and $\sim r$ are $1 - \alpha$ and $1 - \beta$ respectively, and, provided they are independent, the probability of their conjunction is $(1 - \alpha) \times (1 - \beta)$ which works out as $1 - \alpha - \beta + \alpha\beta$; hence the probability of the negation of their conjunction is

$$1 - (1 - \alpha - \beta + \alpha\beta),$$

which works out as

$$\alpha + \beta - \alpha\beta,$$

which is the Disjunction Rule for independent propositions. If q and r are not independent but mutually exclusive, the probability of $q \vee r$ takes the simpler form $\alpha + \beta$. But the proof of this and the unrestricted Disjunction Rule we must leave to the next Chapter.

IV

PROPOSITIONAL FUNCTIONS AND CONDITIONAL PROBABILITIES

IN probability theory we often need to deal with a large number of cases which are qualitatively identical and numerically distinct. We have had, so far, no way of symbolizing propositions about qualitatively identical but numerically distinct cases. It is natural to use the notation of a propositional function to do so. Most experts talk of "events" or "cases". But these two words suffer from the grave disadvantage of not distinguishing between type and instance.† The "event" of the coin coming down heads may refer to the event-instance of *this* coin coming down heads *next* toss, or to the event-type of *a* coin coming down heads on being tossed. In the first sentence quoted from Kolmogorov on p. 27, elementary events are *members of* random events, but the word 'event' obscures the class-membership relation, and suggests misleadingly that it is rather one of class-inclusion. Moreover, the Boolean algebra of events, although feasible, is more artificial than the propositional calculus and the first-order functional calculus. I therefore avoid talking of events and cases, and speak instead of *propositional functions* having probabilities assigned to them. A propositional function is written in the form

$$F(x)$$

where F is a constant predicate denoting a constant property, and x is a variable, standing in lieu of a name denoting a particular object or situation. Thus $F(x)$ might symbolize the propositional function

'   is red'

which would become a particular proposition when the blank was filled by the name of some object, *e.g.*

† For the distinction, and its importance in a neighbouring field, see J. R. Lucas, "Causation", in R. J. Butler, ed., *Analytical Philosophy*, I, Oxford, 1962, pp. 34–7.

'The sofa is red'

which might itself be symbolized by, say,

$$F(g)$$

$F(\ )$, as we said, standing for '   is red', and g standing for the sofa.

With this notation, we can symbolize a number of qualitatively identical but numerically distinct propositions thus:

$$F(g_1), F(g_2), F(g_3), \ldots, F(g_n),$$

the difference in the subscripts of the g expressing the numerical distinctness, the sameness of the g and of the F expressing the qualitative indentity. We can express the same point, slightly more succinctly, by assigning a probability to the propositional function itself,

$$F(x),$$

when the variable x has

$g_1, g_2, g_3, \ldots, g_n$ (if there are only a finite number of g's)

or

$g_1, g_2, g_3, \ldots, g_n, \ldots$ (if there are an infinite number of g's)

as its range of values

Thus, if $g_1, g_2, \ldots$ are successive tosses of a coin, and F is the property of coming down tails, then we can express 'The probability of a coin's coming up tails is 0·471' by giving $F(x)$ the value 0·471 when x ranges over the g's.

Probabilities may properly be assigned to propositional functions, because truth-values can. Although we tend to think of propositional functions as always being quantified, they do not have to be. Indeed, in most formulations of the functional calculus (or predicate calculus, as it is sometimes called), $(x)F(x). \supset F(y)$, where y may be a free variable, is an axiom, and conversely the rule of generalisation enables us to infer $(x)F(x)$ from $F(x)$ if x is a free variable. In view of this axiom and this rule, to assert $F(x)$ is tantamount to asserting $(x)F(x)$, and to assert $\sim F(x)$ is tantamount to asserting $(x) \sim F(x)$. It is because of this, that we tend to forget that propositional functions, by themselves and not lying in the scope of any quantifier, can have truth-values, and hence probabilities too.

We shall use not Whitehead and Russell's first-order functional calculus but T. J. Smiley's many-sorted logic.† Whitehead and Russell make very little use of their individual variables and the universe of discourse, but put all the weight on the predicate variables. The formulae $(x)F(x)$ and $(\exists x)F(x)$ hardly ever occur by themselves; instead $(x).F(x) \supset G(x)$—if we want to say that all F's are G—or $(\exists x).F(x) \,\&\, G(x)$—if we want to say that some F's are G. But we shall need to distinguish being *an F*-sort of thing from being G—having the property G—even though the two are related and the distinction is not always an absolute one. Moreover we want to avoid using "material implication". It is only, so to speak, a logical accident that material implication can be used in two-valued logic to render 'if . . . , then . . .'; it is like using a screwdriver as a chisel to open a packing case—the under-lying subject matter is so coarse-grained that almost any tool will serve to prize the relevant parts apart. As soon as we are dealing with anything more subtle, like a many-valued logic, and more especially with a probability calculus with a non-denumerable infinity of truth-values, material implication leads to absurdity. It is only tolerable, and then only just tolerable, to regard

$$\sim q \lor r$$

as an implication at all, if we are to disregard all values of q and r except the two extreme ones. Else, the rule of *modus ponens* and the definition of implication in terms of other truth-functional connexions pull apart. If we want to have any more sophisticated calculus than the black-or-white two-valued calculus of elementary logic, we must give up material implication and express 'All F-sort things are G' not by

$$(x).F(x) \supset G(x),$$

but by

$$(f)G(f)$$

or, better, by simply asserting the unquantified

$$G(f).$$

From this it is natural to generalise, and give such a propositional function not only the two truth-values True and False, but any other probability-value in between.

---

† Timothy Smiley, "Syllogism and Quantification", *The Journal of Symbolic Logic*, 27, 1962, pp. 58–60.

Propositional functions of the form $G(f)$ clearly contain two effective terms, the individual variable f as well as the predicate variable G. The fundamental reason is that if we are going to have a whole set of propositions which we can count as qualitatively identical though numerically distinct, we must specify the features by which they are to count as qualitatively identical; we must say in advance whether it is swans or ravens we are considering, or tosses of a coin or throws of a die or spins of a roulette-wheel. And this is to use a propositional function in the form $G(f)$. That is what is right in the doctrine of the Cambridge thinkers, W. E. Johnson, Lord Keynes, and Sir Harold Jeffreys, that probability is a relation. In order to specify a mathematical probability we need to indicate *two* types: one the range, or universe of discourse—the type of individual or event being considered, *e.g.*, men, live births, tosses of a coin—the other the property, feature, or characteristic which is possessed by some, but not all, of the instances of the first type, *e.g.* being bald, being male, coming down heads. According to Keynes, Gilman was the first to see this:

Probability has to do, not with individual events, but with classes of events; and not with one class, but with a pair of classes—the one containing, the other contained. The latter being the one with which we are principally concerned, we speak by ellipsis, of its probability without mentioning the containing class; but in reality probability is a ratio, and to define it we must have both correlates given.†

Keynes himself, however, together with Johnson and Jeffreys, misconstrued the two terms as being premiss and conclusion of an argument, and probabilities as attaching to the relation between them. They use the notation a/h or $P(q/p)$ which is read as 'the probability of a on hypothesis h' or 'the probability of proposition q on data p'.‡ Normally, although not always, when we make a numerical probability-judgement we are assigning the probability to a propositional function expressing the occurrence of a certain property, feature, or attribute in a certain range or

† B. I. Gilman, "Operations in Relative Number with Applications to the Theory of Probabilities", in John Hopkins, *Studies in Logic*, Baltimore, 1883; quoted J. M. Keynes, *A Treatise on Probability*, London, 1921, p. 156.

‡ J. M. Keynes, *A Treatise on Probability*, London, 1921, pp. 4, 40, 43–4; Harold Jeffreys, *Theory of Probability*, 3rd ed., Oxford, 1961, p. 20.

universe of discourse. We therefore need to specify both the property and the range before we have said what the propositional function is which we are assigning a probability to. Jeffreys almost says this. He says

It is no more valid to speak of the probability of a proposition without stating the data than it would be to speak of the value $x + y$ for given x, irrespective of the value of y.†

But he uses the word 'data', and elsewhere 'evidence', instead of 'range' or 'universe of discourse' or some such, and it is a crucial difference. The whole Logical Relation theory is vitiated, I believe, because it makes out probability to attach to the relations between evidence and conclusions rather than to propositions, or propositional functions, by themselves. Jeffreys argues that the probability must depend on the evidence

Suppose that I know that Smith is an Englishman, but otherwise know nothing particular about him. He is very likely, on that evidence, to have a blue right eye. But suppose that I am informed that his left eye is brown—the probability is changed completely.‡

But this is not a probability of a singular proposition whose subject is Smith: for we know nothing particular about him, except first that he is an Englishman and then that his left eye is brown. Under the guise of a singular proposition we are really considering some sort of general proposition, first about Englishmen then about Englishmen with brown left eyes. Let us symbolize the universe of discourse constituted by Englishmen by e, and, less felicitously, the universe of discourse constituted by Englishmen with brown left eyes by the composite symbol eb.§ I then say that in Jeffreys' example we are considering the two probabilities

first of the propositional function Blue (e)
and secondly of Blue (eb)

where Blue is the predicate 'has a blue right eye'. The probability of the former is sizeable, but of the latter very small indeed.

The two differences between the Logical Relation theory and the account given here are first that the former accepts that grammatically singular sentences express singular propositions

---

† Harold Jeffreys, *Theory of Probability*, 3rd ed., Oxford, 1961, p. 15.
‡ *Op. cit.*, p. 15.
§ See further below, pp. 59–60.

whereas I claim that in the relevant cases they are covertly universal† and should be regarded as expressing propositional functions; and secondly that the former talks of the probability of a proposition B given certain data E, or relative to certain evidence E, whereas I claim that we can perfectly well talk of the probability of a proposition or propositional function without specifying evidence, although there is a problem of how we specify what the propositional function itself is. The Logical Relation theory has assimilated and confused the two different things: (i) the evidence for a proposition, (ii) the specification of that proposition. If I am concerned with whether an Englishman has a blue right eye, the evidence will consist of various statistics about the occurrence of blue-eyed men in England, together perhaps with some more general considerations supporting the Mendelian theory. Similarly, if I assign a particular probability to the propositional function that a man aged twenty smoking 40 cigarettes a day will die of lung cancer before he is forty, the evidence will again consist of statistics, together perhaps with certain observations on the effect of painting tar on rabbits' ears *etc*. The additional information adduced in Jeffreys' example does not alter our estimate of what the probability of any propositional function *is*, but alters our view of which propositional function is relevant. The relevant fact about Smith is no longer simply that he is an Englishman, but that he is an Englishman with a brown left eye. We therefore no longer apply to Smith's case the propositional function of an Englishman having a blue right eye but the propositional function of an Englishman-whose-left-eye-is-brown having a blue right eye. The additional information has altered the specification of the propositional function to be applied in Smith's case, but has not altered the probability value to be assigned to either of the propositional functions in question. We therefore need to distinguish these two uses of 'information', 'data', or 'evidence'. The one affects our general stock of knowledge, and may lead us to alter the probabilities we have provisionally assigned to some propositional function. The other concerns only which piece of general knowledge is applicable to the particular case. It determines the subject

† Probability-judgements are always covertly universalisable because they claim to be based on reasons; see above Ch. I, p. 5, see also D. H. Mellor, "Chance", *Proceedings of the Aristotelian Society*, Supplementary Volume, XLIII, 1969, §6, p. 31.

of the question we are asking—whether about Englishmen or Englishmen-with-brown-left-eyes—to which the evidence properly so called—statistics of eye colour among Englishmen, *etc.*—will provide the answer. If we are talking only about particular cases—which philosophers and scientists seldom are—we are not committing a solecism in talking of both sorts of information as evidence. Keynes and Jeffreys are not committing a crime against English usage. But there is an obvious difference between these two sorts of information, and it would serve the cause of clarity to confine the word 'evidence' to the former, the facts on which generalisations are or might be made, not the facts which determine what generalisations are or might be applicable to a particular case.

We need to amend Jeffreys' claim "It is no more valid to speak of the probability of a proposition without stating the data than it would be to speak of the value of $x+y$ for given x irrespective of the value of y", and say instead that when we are dealing with a propositional function it makes no sense to speak of its probability specifying only the value of the predicate variable without specifying also the range or universe of discourse of the free individual variable: it makes no more sense to speak of the probability $F(y)$ for given F (*e.g.* Blue) irrespective of the range of y, than it would be to speak of the value of $x+y$ for given x irrespective of the value of y. We need to know not only what is being asked (Jeffreys' x) but what it is being asked about (Jeffreys' y) before we can give an answer—even a probabilistic one—(Jeffreys' z).

Jeffreys supports the Logical Relation theory with other arguments, which are, I shall maintain, essentially subjectivist in import. He argues:

Rule 4. The theory must provide explicitly for the possibility that inferences made by it may turn out to be wrong. A law may contain adjustable parameters, which may be wrongly estimated, or the law itself may be afterwards found to need modification. It is a fact that revision of scientific laws has often been found necessary in order to take account of new information—the relativity and quantum theories providing conspicuous instances—and there is no conclusive reason to suppose any of our laws are final. But we do accept inductive inference in some sense; we have a certain amount of confidence that it will be right in any particular case, though this confidence does not amount to logical certainty. . . .

These five rules are essential. . . . The fourth states the distinction between induction and deduction.†

1. 2. The chief constructive rule is 4. It declares that there is a valid primitive idea expressing the degree of confidence that we may have in a proposition, even though we may not be able to give either a deductive proof or a disproof of it. . . . We need to express its rules. One obvious one (though it is very commonly overlooked) is that it depends both on the proposition considered and on the data in relation to which it is considered. . . . It is a fact that our degrees of confidence in a proposition habitually change when we make new observations or new evidence is communicated to us by somebody else, and this change constitutes the essential feature of all learning from experience. We must therefore be able to express it. Our fundamental idea will not be simply the probability of a proposition p, but the probability of p on data q. Omission to recognize that probability is a function of two arguments, both propositions, is responsible for a large number of serious mistakes.‡

Jeffreys' rule 4 is a sophisticated version of the argument that a probability statement may have been made entirely reasonably and yet turn out false in the event.§ If I say 'It probably will rain tomorrow', I may be speaking entirely reasonably, having had recourse to the best available meteorological information and evidence, and making proper calculations on that basis: and it would be hard then to say that I was wrong when I said 'It probably will rain tomorrow.' And yet it could be the case, even in England in November, that when tomorrow came, it did not rain at all. And so we may feel inclined to say that both 'It will probably rain tomorrow' said on Tuesday, and 'It has not rained all day today' said late on Wednesday, are together true. And we resolve this apparent paradox by saying that the word 'probably' is elliptical, and means really 'probably-on-the-basis-of-information-at-present-in-my-possession'.

We have seen why this answer will not do. If 'probably' and all related words were elliptical, as alleged, and were short for 'probably-on-the-basis-of-information-at-present-in-my-possession' then there would be no inconsistency at all between my saying 'It probably will rain' and your saying 'It probably will not'. We should be talking about different relations, just as if I

† *Op. cit.*, pp. 8–9.   ‡ *Op. cit.*, p. 15.
§ See above, Ch. I, p. 3, and Ch. II, p. 17.

were saying 'It is a long way from home' and you were saying 'It is not a long way from home'. Although the word 'probably' does hedge, and does give warning that one may turn out to be wrong, it does not hedge so much that no claim whatever is being made. If I say 'It probably will rain tomorrow', I am hedging, and am warning you not to place too much reliance on my prediction, not to treat it as gospel truth: but I am none the less saying something, however tentatively: and if you, however tentatively, disagree, we are disagreeing about something. We are, in fact, talking about the same thing, namely whether it will rain tomorrow or not, and not two different things, namely the relation between the proposition that it will rain on Wednesday and my other beliefs, and the relation between that same proposition and your other beliefs. Our respective beliefs are the basis on which we put forward our respective probability statements, but are not part of them. The evidence and arguments which I can adduce in support of my tentative prediction 'It will probably rain tomorrow' are my reasons for saying what I am saying but do not constitute part of what I am saying: they are criteria for the correct use of the sentence, not part of its meaning. If you contest my probability statement, I may adduce my arguments and facts, and you may adduce yours, and we each shall throw into the common pool such relevant considerations and relevant pieces of information at our disposal. This may persuade either or both of us to change his assessment of probability: but if so, one or both of us will have changed his opinion, not changed the subject; he will have changed the assessment he had made of the probability of its raining on Wednesday, but it will still be the same topic of conversation, namely the probability of its raining on Wednesday. If this were not so, we should have to maintain that arguments about probability were peculiarly fruitless; they necessarily could never get anywhere, because either both the parties stuck to their initial position, so that they were just where they were when they started, or if one of them changed his mind as a consequence of the argument, it would mean that he had, unwittingly, changed the subject. "Either I go on saying the same as I was saying or let's talk about something else."

For these reasons we cannot allow that 'probably' is elliptical, and means really 'probably-on-the-basis-of-the-information-at-present-in-my-possession'. If I said on Tuesday that it would rain on Wednesday, and it did not rain on Wednesday, I may well have been entirely justified on Tuesday in saying what I did, and not at all to blame for the non-fulfilment of my cautious prognostication. In this sense it was not wrong to say that it would probably rain on Wednesday. But I cannot pretend that what I said was *true*. It was reasonable. It was quite right for me to say it in the circumstances; but unfortunately my prediction, hedged as it was, did not come true. How lucky I hedged. But I cannot claim that my hedging makes my prediction invulnerable to subsequent falsification, for, if that was so, I was not making any prediction whatsoever, and might as well have kept my mouth shut. It is only by sticking my neck out, however well shielded, that I succeed in saying anything about reality. If I feel too unsure of my judgement to risk saying simply 'It will rain tomorrow' I can hedge by adding 'probably', and if that is insufficient hedge I can add 'in my opinion, for what it is worth', and if I feel that I am still being unduly dogmatic I can add 'but do not take it from me; you must make up your mind for yourself'. But when I have put in every conceivable hedge and disavowal of infallibility, I have still said something, which, if taken at its face value, would incline my hearer more to the view that it was going to rain than to the view that it was not going to rain. If I want to go further than this, and say even less, so that there will be no possibility of blaming me if it does not, in the event, rain, the correct linguistic procedure is not to pick on 'probably', 'possibly', 'perhaps', after saying 'It will rain tomorrow', but to add Humpty Dumpty's rider 'or else it will not'.

Jeffreys wants a Humpty Dumpty all-ways protection. "The theory must provide explicitly for the possibility that inferences made by it may turn out to be wrong." Without numbering myself among the Popperian elect, I would say, with Popper, that this is just what a theory ought not explicitly to provide for.† "A law", Jeffreys continues, "may contain adjustable parameters, which may be wrongly estimated, or the law itself may be afterwards found to need modification." Of course a law *may* do either of these, but it is no virtue to do so. Theories may lead to wrong inferences and laws may have to be adjusted, but the point of inferences, even probabilistic ones, is that they should be right,

† K. R. Popper, *The Logic of Scientific Discovery*, London, 1959.
5—C.O.P.

and the less adjusting required *ex post facto*, the better. Inductive arguments, unlike deductive ones, can go wrong, in the sense of yielding conclusions which turn out to be false, and Jeffreys is quite right to seize on this as "the distinction between induction and deduction". But the whole point of this is that the inference went wrong in that the conclusion was false. We must not then cover ourselves by saying that we were not talking about the conclusion but the logical relation between the premisses and the conclusion. Because then we shall be back in an entirely deductive world, like Carnap's, in which nothing can ever by mischance go wrong, and all our probabilities are completely insulated from cruel facts.

The Logical Relation theory is in unstable equilibrium. Although its proponents are not subjectivists, some of their arguments lead to subjectivism. Although they are concerned to elucidate the probability that satisfies the traditional calculus, they often seem to be talking not of probability theory but of confirmation theory; which, if it ever is developed satisfactorily will be very different from traditional probability theory: and while there is a locution in which we talk of arguments rather than propositions being probable, the normal locution, as Keynes himself admits,† is to ascribe probability, like truth, to the conclusion of the argument rather than to the argument itself. If we reject subjectivism, and do not want to study confirmation theory, we shall not want to accept Jeffreys' arguments for construing all the information we use in reaching a probability-judgement as *data* or evidence; and then it will be natural to distinguish some information, which really is evidence, from what is not so much evidence as part of the specification of the relative propositional function. Our probability-judgements depend on both; but in different ways. Evidence, properly so called, affects the answer in a straightforward way: but the specification of the propositional function affects it much more radically, because it determines not simply the answer we give, but the question we ask, to which the answer we give is an answer.

Sometimes, but only sometimes, different universes of discourse are related to one another in such a way that the probabilities in the one determine those in the other; so that although we alter our question, and ask about the probability of a different

† *Op. cit.*, p. 4.

propositional function in a different universe of discourse, yet the answer to the new question is already determined by answers already given to previous ones. It is only possible in special cases. We can go from one universe of discourse, say that of the f's, to a larger, say that of the g's, provided the former is a subclass of the latter; that is to say, all the f's are g's. And, conversely, we can go from a larger universe of discourse, say that of the g's, to a smaller, say that of the f's, provided the latter is a subclass of the former; that is to say, all the the f's are g's. But only in these two cases, where one universe of discourse is entirely included in the other, can probability theory accommodate the transition from one to the other. If the two universes of discourse merely overlap, without either being entirely included in the other, there is no systematic relation between the probabilities of propositional functions in the one and of those in the other: just as, more obviously, there is no relation between propositional functions ranging in universes of discourse that are totally exclusive. The probability of a newly-born elephant living to be seventy is one thing, and that of a newly-born man is another: but there is no relation between them, because elephants and men are different. Similarly, though less obviously, the probability of a woman having fair hair and that of a native of England having fair hair are unrelated, because, although some women are natives of England, many are not, and many natives of England are not women. Only when one universe of discourse is entirely included in the other will the probabilities of propositional functions ranging in the one be systematically related to the probabilities of propositional functions ranging in the other. In these cases the universe of discourse has, in effect, undergone a magnification or, conversely, a reduction, and a systematic compensatory reduction, or magnification, is all that is called for.

Smiley's many-sorted logic enables us to handle these changes of universe of discourse. In it, as we have already seen,† the universe of discourse is indicated by our choice of small italic letter to denote the individual variable, and we use the corresponding capital letter to express the corresponding feature. Thus $H(g)$ represents the propositional function that a g—i.e. an individual in the universe of discourse of G-type things—has feature H, while $G(f)$ represents the propositional function that an f—i.e.

† p. 48.

an individual in the universe of discourse of F-type things—has feature G. The correlation between capital and small letters provides us with alternative ways of symbolizing quantified propositions. We can either use the individual variables, and say $(g)H(g)$, $(\exists g)H(g)$ etc., or we can follow Whitehead and Russell, and make the individual variables vacuous by having a large universe of discourse, and say $(x).G(x) \supset H(x)$, $(\exists x).G(x) \mathbin{\&} H(x)$ etc., or, provided f ranges over a large enough universe of discourse, $(f).G(f) \supset H(f)$, $(\exists f).G(f) \mathbin{\&} H(f)$, etc. We thus need rules to connect the individual variables f, g, h, etc. with the corresponding predicate variables F, G, H, etc. Smiley uses the following axiom schemes and rules, in which ϕ, ψ, and $\phi(f)$, are any well-formed formulae, f and g are any variables, $\phi(g)$ has free occurrences of g wherever $\phi(f)$ has free occurrences of f, and F is the sortal predicate corresponding to the variable f:†

A1. Axioms for the propositional calculus.
A2. $(f)(\phi \supset \psi) \supset .\phi \supset (f)\psi$, if f is not free in ϕ.
A4.‡ $(f)\phi(f) \supset .F(g) \supset \phi(g)$.
A5. $F(f)$.

Rule of generalisation: from ϕ infer $(f)\phi$.
Rule of detachment: from ϕ and $\phi \supset \psi$ infer ψ.

A4 and A5 are the special axioms connecting the individual variables with their corresponding predicates. A4 in Smiley's presentation contains a quantifier; we want to avoid quantifiers, and concern ourselves exclusively with free variables. If we had the premisses $\phi(f)$ and $F(g)$ in Smiley's system we should still be able to obtain $\phi(g)$, because we could first use the rule of generalisation to obtain $(f)\phi(f)$, and then use A4. This derivation can be expressed briefly

$$\phi(f), F(g) \vdash \phi(g)$$

† I have altered Smiley's a to f, and b to g, because since the letter a is an English word, and the spoken b sounds like an English word, they might cause confusion in the discussion which ensues.

‡ A3, introduced earlier in Smiley's presentation, is redundant after he has added A4 and A5.

where the sign ⊢ can be read 'yields'.† This result expresses the rule of the syllogism for propositional functions—if a g is F and an f is ϕ, then a g is ϕ. We shall need it later.‡

We shall be concerned only with universes of discourse, or "sorts", such that one is entirely included in the other; e.g. with the universe of the f's, where every f is also a g. To make it clearer what is meant, I have introduced two further, somewhat unconventional, symbolisms. I shall use the joined symbols, fg, gf, etc., as individual variables ranging over the universe of discourse of all those things that are both g's and f's, etc.; I shall also use the joined symbols FG, GF, etc., as predicate variables indicating the possession both of feature F and of feature G, etc.§; and it is convenient to write $\bar{H}(g)$ as an abbreviation for $\sim H(g)$. Thus $H(fg)$ symbolizes the propositional function that an f that is also a g possesses the feature H; and $HG(f)$ symbolizes the propositional function that an f possesses both the feature H and the feature G. We could represent these without the use of a special symbolism: but the condensed notation is perspicuous. We should also note the possibility of having more than one variable ranging over a given universe of discourse. Besides $F(g)$ we can have $F(g')$, $F(g'')$, etc., with g', g'', etc., ranging over all the g's.

The symbolism of the small italic letters representing individual variables ranging over specified universes of discourse needs to be handled with care. One might suppose that since we can conjoin variables, writing fg for an-f-which-is-also-a-g, or eb for an Englishman-with-a-brown-left-eye, we can also disjoin them or negate them. But this is not so. Aristotle observed that we cannot negate the subject;¶ $\bar{f}$—a non-f—is meaningless, unless understood as embedded in some suitable universe of discourse. If f is a circle, and provided I can restrict myself to the universe of discourse of geometrical figures, I can intelligibly talk of non-circles, and say that some are triangles, and not all are convex. But where the proviso does not hold, the term non-circle ceases to mean

† For a treatment of the ⊢ sign, see Alonzo Church, *An Introduction to Mathematical Logic*, Vol. I, Princeton, 1956, Ch. I, §13, and note 181.

‡ On p. 63.

§ Since conjunction is commutative, fg is the same as gf, and FG is the same as GF.

¶ *Categories*, 3b24. Ὑπάρχει δὲ ταῖς οὐσίαις καὶ τὸ μηδὲν αὐταῖς ἐναντίον εἶναι. It is characteristic of substances that there is nothing contrary to them.

anything. It cannot refer to anything-which-is-not-a-circle, because there is no criterion of identity for 'anything'. It is true of most men that they are not circular, and it is not true of any college that it is circular, but we should feel a justifiable disquiet in counting among our anythings-which-are-not-circles *both* normally shaped men *and* the colleges composed of them. Circles can be referred to, but the disparate *congerie* of triangles, colleges, colours, sounds, numbers, chemical substances, smells, concepts, propositions, Old Uncle Tom Cobley and all, is not a possible subject of discourse, and cannot constitute a universe of discourse or be referred to by a subject term. Even with predicates we are chary of talking too freely of being non-circular or non-prime, and restrict the negated predicate to the same universe of discourse as the predicate itself. Geometrical figures can be circular or non-circular, and natural numbers can be prime or composite: but the colour green, we say, is neither circular nor non-circular, and π neither is, nor is not, prime. It is too easily assumed that predicates obey a Boolean algebra without any provisos or restrictions. We shall later argue against this assumption in another context:† here we need only note that the subject term is even less amenable to negation than the predicate; nonentities are much more unreal than negative properties. We therefore do not admit the term $\bar{f}$ by itself, although we do admit terms such as $g\bar{f}$ or $f\bar{g}$, since we can make sense of a g-which-is-not-an-f, or an-f-which-is-not-a-g, because in these cases there is a universe of discourse intelligibly specified by the positive term which the negative term merely specifies further.‡

We can express in the symbolism of many-sorted logic the rules for negation and conjunction we have already derived.

   (1) If α is the probability of $G(f)$, then $1-\alpha$ is the probability of $\sim G(f)$.

   (2) If α is the probability of $G(f)$, and β is the probability of $H(f)$, then $\alpha \times \beta$ is the probability of $GH(f)$, provided $G(f)$ and $H(f)$ are independent.

† In Ch. V, pp. 86–94.

‡ It is usually convenient to put the negative, merely specifying, terms first, and the most general positive one, the subject term, last. This is contrary to English word order; but so is our symbolization of propositional functions as $G(f)$ etc. In Ch. VIII, however, the order is reversed for typographical reasons.

Thus we can apply the calculus of probabilities to the predicates, F, G, H, ranging in any one universe of discourse, just as before. What we need to determine now is the effect of changing the universe of discourse, using the fact that we can view a universe of discourse either as the range of the appropriate variable or as a proper subclass of some larger universe of discourse, specified as those members of that larger universe that have the further feature specified. We can either talk of the gf's—the-g's-which-are-also-f's—or of the fg's—the-f's-which-are-also-g's—or talk, instead, of those g's of which $F(g)$ is true.

Let us consider—at first only intuitively—the case where the universe of discourse is reduced. That is, instead of discussing the g's, and the propositional function that a g is H, we discuss the fg's and the propositional function that an fg is H. We can write the former $H(g)$ and the latter $H(fg)$. If we reduce the universe of discourse from g's to the fg's, how will the probabilities of the propositional functions ranging among the g's be affected? First, we shall be limited to those propositional functions which can be represented in the g-type discourse as a conjunction, a conjunction of the original propositional function $H(g)$, that g is H, with the propositional function $F(g)$, that g is F, i.e. that g is an f. That is, we shall not be concerned simply with the propositional function $H(g)$, but with the conjunct propositional function $HF(g)$.† For we are limited to those g's which are also f's, that is, to those g's of which $F(g)$ is true. Those g's of which $F(g)$ is not true do not concern us, even though $H(g)$ may be true: the propositional function $H\overline{F}(g)$—i.e. $(H(g)$ & $\sim F(g))$— is to be discounted, and only $HF(g)$ counted. The same argument holds for all other propositional functions. Reducing the universe of discourse from the g's to the fg's is tantamount to replacing every propositional function, $H(g)$, $K(g)$, etc., by the conjunction $HF(g)$, $KF(g)$, etc. What is the effect of this? We might at first suppose that the probability of $HF(g)$ would be given by the simple Conjunction Rule, and be the product of the probabilities of $H(g)$ and $F(g)$ respectively. But that would be to forget the requirement of independence. The simple Conjunction Rule applies only if the propositional functions are independent of each other. But this we cannot always assume. Often the possession of feature F does have some relevance to the possession of feature H—were it not so, the

† *i.e.* $H(g)$ & $F(g)$; see above, p. 59.

narrowing down of the universe of discourse from all the g's to only those g's which were also F would make no difference to the probability of possessing the feature H. It is just because the sub-class of fg's among the g's needs, in general, a different assignment of probabilities to propositional functions, that there is a problem. But then, if there is a problem, the propositional functions $H(g)$ and $F(g)$ are, in general, not independent, and the probability of their conjunction is *not* given by the simple, restricted Conjunction Rule.

Nevertheless, our intuition to apply the product rule was correct. It was only wrong in assuming that the effect of considering the conjunction $HF(g)$ in place of $H(g)$ was always to multiply the probability of $H(g)$ by that of $F(g)$. In truth, we do not know what the probability of $F(g)$ and $H(g)$ will, in general, be. It will depend on the bearing that $F(g)$ has on $H(g)$ for each particular H. All we can say is that whatever the probability of the propositional function $HF(g)$, it will be a normal propositional function ranging over g, and subject to the usual rules of the probability calculus. In particular, the conjunction of the two conjunct propositional functions $HF(g)$ and $KF(g)$ will be $HKF(g)$; and the propositional function $\bar{H}F(g)$ will behave like the complement of $HF(g)$ within $F(g)$. Therefore, if we confine our attention, within the universe of discourse of the g's, to those propositional functions of the form $HF(g)$, $KF(g)$, *etc.*, we shall find that in general their probability-*values* are different from those of $H(g)$, $K(g)$, *etc.*, but the *rules* for conjunction are exactly the same, and for negation nearly the same. The only difference is that while for negation normally, if α is the probability of $H(g)$ then $1 - \alpha$ is the probability of $\sim H(g)$, for negation subject to the condition F, where γ is the probability of $F(g)$, then if α is the probability of $HF(g)$, $\gamma - \alpha$ is the probability of $\bar{H}F(g)$. The effect of considering the propositional functions $HF(g)$ and $\bar{H}F(g)$, instead of $H(g)$ and $\sim H(g)$ simply, has been to scale down the *total* probability, assigned to $HF(g) \vee \bar{H}F(g)$ by a factor of γ, although it does not, in general, scale down the separate probabilities to be assigned to $HF(g)$ and to $\bar{H}F(g)$ by γ in each case alike. If we were to scale up the probabilities of $HF(g)$, $KF(g)$, *etc.* by a factor of $1/\gamma$, we should have these conjunct propositional functions with probabilities that satisfied all the requirements of the probability calculus. By changing our universe of discourse from the g's to the fg's we have limited our concern to those propositional functions ranging over the g's that had the form $HF(g)$, $KF(g)$, *etc.*; these propositional functions must obey all the rules of the probability calculus—else the universe of discourse of the g's would be incoherent. But, considered as propositional functions ranging over the g's, their total probability will amount to $\text{Prob}[F(g)]$, which is γ, instead of to $\text{Prob}[G(g)]$, which is 1. We therefore need to mark the transition by "re-normalising", that is, by scaling up by a factor $1/\text{Prob}[F(g)] = 1/\gamma$.

This, therefore, is what we do when we view the propositional functions $HF(g)$, $KF(g)$, *etc.*, not as a proper subclass of all the propositional functions ranging over the universe of discourse of the g's, but as *all* the propositional functions there are that range over the universe of discourse of the fg's. For, using the syllogism rule we obtained from Smiley's A4, $G(fg)$, $HF(g) \vdash HF(fg)$. But $G(fg)$ is given, since all fg's are g's; and $HF(fg) \vdash H(fg)$; so $HF(g) \vdash H(fg)$. That is, if $HF(g)$ holds in the universe of discourse of the g's, then $H(fg)$ holds in the universe of discourse of the fg's. Conversely, if $H(fg)$ holds in the universe of discourse of the fg's, then for every such fg, $FG(fg)$ holds by A5, and hence $F(fg)$, and so $HF(fg)$; but all these fg's are g's-which-are-F, and therefore of *them* (*i.e.* those g's which are F) it is true that $HF(g)$ holds. That is $H(fg)$, $F(g) \vdash HF(g)$. And so $HF(g)$ holds whenever $H(fg)$ and $F(g)$ both do.

So much by way of intuitive exploration. I now give a more formal argument. The first step is to establish the equivalence of $HF(g)$ and $H(fg) \& F(g)$ in Smiley's many-sorted Logic. If $HF(g)$ holds, then $H(fg) \& F(g)$ holds. For

(i) $\qquad\qquad HF(g)$, $G(fg) \vdash HF(fg)$ Syllogism rule (see p. 58 above, substituting HF for ϕ, g for f, G for F, and fg for g)

(ii) $\therefore$ $\qquad HF(g)$, $G(fg) \vdash H(fg)$ definition of HF

(iii) but $\qquad\qquad \vdash FG(fg)$ Axiom 5

(iv) $\therefore$ $\qquad\qquad \vdash G(fg)$ definition of FG

(v) $\therefore$ $\qquad HF(g) \vdash H(fg)$ from (ii) and (iv).

(vi) Also $\quad HF(g) \vdash F(g)$ definition of HF

(vii) $\therefore$ $\qquad HF(g) \vdash H(fg) \& F(g)$ from (v) and (vi).

Conversely, if $H(fg)\,\&\,F(g)$ holds, then $HF(g)$ holds. For

(i)	$F(g),\ G(fg) \vdash F(fg)$	Syllogism rule (see p. 58 above, substituting F for ϕ, g for f, G for F, and fg for g)
(ii) but	$\vdash G(fg)$	as (iv) on previous page
(iii) $\therefore$	$H(fg)\,\&\,F(g) \vdash HF(fg)$	from (i) and definition of HF.
(iv) Now	$\vdash G(g)$	Axiom 5
(v) $\therefore$	$F(g) \vdash FG(g)$	from (iv) and definition of FG.
(vi) Also	$HF(fg),\ FG(g) \vdash HF(g)$	Syllogism rule (see p. 58 above, substituting HF for ϕ, fg for f, and FG for F)
(vii) $\therefore$	$HF(fg),\ F(g) \vdash HF(g)$	from (v) and (vi)
(viii) $\therefore$	$H(fg)\,\&\,F(g) \vdash HF(g)$	from (iii) and (vii).

Hence the propositional function $HF(g)$ is equivalent to the conjunction of the propositional functions $H(fg)\ \&\ F(g)$. Once we have established this, we can argue that the probability assigned to $HF(g)$ must be the same as that assigned to $H(fg)\ \&\ F(g)$, and must depend on the probabilities assigned to $H(fg)$ and $F(g)$ separately; and this dependency must satisfy the six conditions given in Chapter III,† namely:

(i) There are no probabilities greater than that of a true propositional function or less than that of a false propositional function.

(ii) As the probability assigned to $H(fg)$ approaches that assigned to $K(fg)$, so must the probability assigned to $HF(g)$ approach that assigned to $KF(g)$.‡

(iii) If the probability assigned to $H(fg)$ is greater than that assigned to $K(fg)$ then the probability assigned to $HF(g)$

† pp. 34–5.

‡ We should note that a similar condition can no longer be stated for the other limb of the conjunction, $F(g)$; for any change on the F would require a change in the f's also, and thus alter the universe of discourse altogether. The corresponding requirement would be: As the probability assigned to $F(g)$ approaches that assigned to $L(g)$, so must that assigned to $HF(g)$ approach that assigned to $HL(g)$—but there is no reason to suppose that the probability of $H(fg)$ is the same as that of $H(lg)$.

must be greater than that assigned to $KF(g)$, except when $F(g)$ is false: or, in symbols, If $\text{pr}[H(fg)] > \text{pr}[K(fg)]$, $\text{pr}[HF(g)] > \text{pr}[KF(g)]$, unless $F(g)$ is false.

(iv) $\text{pr}[F(g)\&(H(fg)\&K(hfg))] = \text{pr}[KHF(g)] =$
$$\text{pr}[FH(g) \& K(fhg)]$$
$$= \text{pr}[(F(g) \& H(fg))\&K(fgh)].$$

(v) If $F(g)$ is true, $\text{pr}[HF(g)] = \text{pr}[(Hg)]$.

(vi) If $F(g)$ is false, $\text{pr}[HF(g)] = \text{pr}[F(g)]$.

It therefore can be argued, exactly as before, that any functions satisfying these conditions can be reduced to a product by regraduating the probabilities. Hence we can reduce the Conjunction Rule to a product rule as before; only, now there is no need to stipulate independence. Whether or not $H(g)$ and $F(g)$ are independent, the probability of $HF(g)$ must depend on $F(g)$ and $H(fg)$. For even if $H(g)$ is not independent of $F(g)$, we have put a value on the degree to which $H(g)$ depends on F by asking not for the probability of $H(g)$ but for that of $H(fg)$. If in fact $H(g)$ is independent of F, then $H(fg)$ will have just the same probability assigned to it as $H(g)$. This will be a special case of the more general one where F may well be a relevant factor, and therefore is taken care of by being asked about in the question 'What is the probability of a thing which is both an f and a g being H?' We can therefore restate the Conjunction Rule without any restrictions, that for any propositional functions, $F(g)$, $H(fg)$ and $HF(g)$

$$\text{Prob}[HF(g)] = \text{Prob}[H(fg)] \times \text{Prob}[F(g)].$$

The first factor on the right-hand side of the equation, $\text{Prob}[H(fg)]$ is often called the Conditional Probability of H given F; it is sometimes written Prob $H|F$ or $P(H|F)$. Kneale† writes in all the factors but inverts the order: he writes $P(\alpha, \beta)$ where I write $\text{Prob}[F(g)]$, and correspondingly $P(\alpha\beta, \gamma)$ where others would have 'γ on condition β'. Kolmogorov‡ defines the conditional probability of the event H under the condition F, written $P_F(H)$, as the quotient $P(HF)/P(H)$. Our argument provides the rationale for such a definition.

If we compare the unrestricted Conjunction Rule with the restricted Conjunction Rule we proved in Chapter III for

† W. C. Kneale, *Probability and Induction*, Oxford, 1949, §25, p. 118.

‡ A. N. Kolmogorov, *Foundations of the Theory of Probability*, tr. N. Morrison, New York, 1956, Ch. I, §4, p. 6.

independent propositions, we see that the latter is a special case of the former with

$$\text{Prob}[H(fg)] \quad \text{equal to} \quad \text{Prob}[H(g)].$$

As we said in Chapter III,[†] we can define independence in terms of the applicability of the restricted Conjunction Rule rather than *vice versa*. We can now express this felicitously for propositional functions by saying that H and F are *independent* of each other within the universe of discourse of the g's if and only if $\text{Prob}[H(fg)] = \text{Prob}[H(g)]$. In such a case, knowing that a g was also an f would not affect the probability of its being H; being an f is irrelevant to whether a g is H or not. We should note that we cannot automatically generalise: pairs of propositional functions may be independent of each other without their all being independent of one another.[‡] Theories have been developed to accommodate various degrees of partial dependence. But they do not concern us here.

In the unrestricted Conjunction Rule the two conjoined propositional functions are no longer *pari passu*. One, $F(g)$, is simple and has a straightforward probability: the other, $H(fg)$, is complex, and therefore has been described as having a conditional probability. But the *rôles* can be reversed. $HF(g)$ is simply shorthand for $H(g)$ & $F(g)$, which, by the ordinary commutative law of the propositional calculus, is equivalent to $F(g)$ & $H(g)$, and could be written $FH(g)$ equally well, and not only could be so written but could be shown to have as its probability the product of the probabilities of $H(g)$ and $F(hg)$. We can thus apply the Conjunction Rule in two different ways to obtain the two different results:

$$\begin{aligned}
\text{Prob}[HF(g)] &= \text{Prob}[H(fg)] \times \text{Prob}[F(g)] \\
&= \text{Prob}[F(hg)] \times \text{Prob}[H(g)]
\end{aligned}$$

Thomas Bayes expressed these results in another form. Equat-

† p. 30.

‡ See A. N. Kolmogorov, *op. cit.*, p. 11, n. 12, for an example, due to S. N. Bernstein; or W. N. Feller, *An Introduction to Probability Theory and its Applications*, Vol. I, 2nd ed., New York, 1957, p. 116.

ing the two right-hand sides of the equations above, and dividing through by $\text{Prob}[F(g)]$, we have BAYES' RULE

$$\text{Prob}[H(fg)] = \text{Prob}[F(hg)] \times \frac{\text{Prob}[H(g)]}{\text{Prob}[F(g)]}.$$

We shall return to this rule and Bayes' two Inversion Theorems which follow from it in Chapter VIII.

The unrestricted Conjunction Rule gives us a new approach to justifying the Negation Rule, and enables us to establish the restricted Disjunction Rule, as applied to exclusive alternatives.

Our previous argument[†] for the Negation Rule required the five assumptions:

   (i) $\text{Neg}(x)$ is continuous and differentiable in the interval $[0, 1]$;
   (ii) $\text{Neg}[\text{Neg}(x)] = x$;
   (iii) $\text{Neg}(0) = 1$;
   (iv) $\text{Neg}'(x)$ is never positive;
   (v) $\text{Neg}'(x) = \text{Neg}'[\text{Neg}(x)]$,

where

$$\text{Neg}'(x) = \frac{d}{dx}\,\text{Neg}(x).$$

We can adapt an argument of Cox's[‡] to replace these two assumptions by *either* a stronger version of (iv), namely

(iv') $\text{Neg}'(x)$ is always negative

or an assumption about Equiprobability, namely

$x = \text{Neg}(x)$ if and only if $x = \frac{1}{2}$,

provided also that (i) is strengthened to stipulate that $\text{Neg}(x)$ should have a second-order derivative.

Cox shows that if we have the unrestricted Conjunction Rule, we can prove

$$\text{Neg}\left[\frac{\text{Prob}[HF(g)]}{\text{Prob}[F(g)]}\right] \times \text{Prob}[F(g)]$$

$$= \text{Neg}\left[\frac{\text{Neg}\,\text{Prob}[F(g)]}{\text{Neg}\,\text{Prob}[HF(g)]}\right] \times \text{Neg}\,\text{Prob}[HF(g)]$$

† Ch. III, pp. 44–5.
‡ R. T. Cox, "Probability, Frequency and Reasonable Expectation", *American Journal of Physics*, Vol. XIV, 1946, pp. 7–8, 12–13; and *The Algebra of Probable Inference*, Baltimore, 1961, Ch. I, pp. 20–2.

for any $F(g)$ and $HF(g)$. If we assign the probability x to $F(g)$ and y to $\sim HF(g)$ we can express this equation as

$$\text{Neg}\left[\frac{\text{Neg}(y)}{x}\right] \times x = \text{Neg}\left[\frac{\text{Neg}(x)}{y}\right] \times y.$$

If we write $\text{Neg}(y)/x$ as u and $\text{Neg}(x)/y$ as v we have

$$x\,\text{Neg}(u) = y\,\text{Neg}(v).$$

Cox then differentiates with respect to x, to y, and to x and y, and eliminates x and y, and then $\text{Neg}'(x)$ and $\text{Neg}'(y)$ to obtain

$$\frac{u\,\text{Neg}''(u)\text{Neg}(u)}{[u\,\text{Neg}'(u) - \text{Neg}(u)]\text{Neg}'(u)} = \frac{v\,\text{Neg}''(v)\text{Neg}(v)}{[v\,\text{Neg}'(v) - \text{Neg}(v)]\text{Neg}'(v)}.$$

In this equation both sides are the same function of u and v respectively. But u and v have been defined in terms of *two* independent variables, x and y, and are therefore themselves independent. The function therefore must be a constant function, say k. But if $\text{Neg}'(u)$ is always negative, this constant must be zero. For when H is the same as $\bar{F}$, $HF(g)$ is false, $\sim HF(g)$ is true, and so y has the value 1, $\text{Neg}(y) = 0$, and hence $u = 0$, and $\text{Neg}(u) = 1$. The denominator of the function becomes $-\text{Neg}'(u)$, which by assumption (iii)' is not zero, while the numerator is zero, since u is. Hence $k = 0$. But neither u nor $\text{Neg}(u)$ in the numerator are always zero, hence $\text{Neg}''(u)$ must be. Hence, integrating twice,

$$\text{Neg}(u) = Au + B$$

and feeding in the special values given by assumption (iii)—*viz.* $\text{Neg}(0) = 1$ and hence (*via* (ii)) $\text{Neg}(1) = 0$—we obtain the canonical Negation Rule

$$\text{Neg}(u) = 1 - u.$$

Assumption (iv'), however, may be questioned. While it would be counter-intuitive for $\text{Neg}'(x)$ ever to be positive, it might perhaps become zero at the end-points 0 and 1, which would be enough to invalidate our present argument that k, and therefore $\text{Neg}''(u)$ must be zero. In that case, we should have to follow Cox's argument through, and obtain the general solution consistent with the boundary conditions at 0 and 1

$$(\text{Neg}(u))^{1-k} = 1 - u^{1-k}.$$

Cox himself argues that one can now take k to be zero by a suitable regraduation of probabilities. But we cannot do this, because we have already regraduated once in order to secure the Conjunction Rule in its canonical form: and if we were to regraduate again we might distort it. We need some further assumption. That of Equiprobability would be adequate. For if we had

$$\text{Neg}(\tfrac{1}{2}) = \tfrac{1}{2}$$

we should have also

$$(\tfrac{1}{2})^{1-k} = 1 - (\tfrac{1}{2})^{1-k},$$

whence $1 + 1 = 2^{1-k}$ so that

$$k = 0.$$

In whatever way we establish the Negation Rule, it will give us the restricted Rule for Exclusive Disjunction. It clearly holds good in the special case of Exclusive and Exhaustive Disjunction; that is,

$$\text{Prob}[q \vee \sim q] = \text{Prob}[q] + \text{Prob}[\sim q],$$

since $q \vee \sim q$ is true, with probability 1, and $\text{Prob}[\sim q] = 1 - \text{Prob}[q]$.

In order to apply this to the general case we need merely to narrow down the universe of discourse until the exclusive alternatives are also exhaustive. If we have $F(g)$ and $H(g)$ as mutually exclusive but not jointly exhaustive in the universe of all G-type things we construct the predicate $K(g)$ which is to be by definition $F(g) \vee H(g)$. Then since F and H are mutually exclusive, $F(kg)$ if and only if $\sim H(kg)$. Hence within the universe of KG-type things,

$$\text{Prob}[F(kg)] = \text{Prob}[\sim H(kg)] = 1 - \text{Prob}[H(kg)]$$

or

$$\text{Prob}[F(kg)] + \text{Prob}[H(kg)] = 1.$$

But by the Conjunction Rule

$$\text{Prob}[F(kg)] = \frac{\text{Prob}[FK(g)]}{\text{Prob}[K(g)]}$$

and

$$\text{Prob}[H(kg)] = \frac{\text{Prob}[HK(g)]}{\text{Prob}[K(g)]}$$

Now, by the definition of K,

$$\text{Prob}[FK(g)] = \text{Prob}[F(g) \& (F(g) \lor H(g))] = \text{Prob}[F(g)]$$

since F and H are mutually exclusive. Similarly

$$\text{Prob}[HK(g)] = \text{Prob}[H(g)]$$

Substituting these values in the equation above, we have

$$\frac{\text{Prob}[F(g)]}{\text{Prob}[K(g)]} + \frac{\text{Prob}[H(g)]}{\text{Prob}[K(g)]} = 1$$

whence

$$\text{Prob}[F(g)] + \text{Prob}[H(g)] = \text{Prob}[K(g)],$$

i.e.

$$\text{Prob}[F(g) \lor H(g)] = \text{Prob}[F(g)] + \text{Prob}[H(g)],$$

which is the restricted rule for Exclusive Disjunction.

We have been able to prove the restricted Disjunction Rule only for propositional functions, whereas we proved the restricted Conjunction Rule for propositions too. It is not surprising. The restriction in the case of the Conjunction Rule was that the propositions should be completely independent, and have nothing at all to do with each other: the restriction in the case of the Disjunction Rule is that they should not be completely independent but should exclude each other. But propositions cannot exclude each other unless they are about the same topic: if I am to contradict what you are saying, I must be talking about the same thing even though I am saying something different about it. Else, I should merely be talking irrelevantly. If we are to disagree we must at least agree on the subject of our disagreement, or we shall not be contradicting one another but merely talking at cross-purposes. Thus two propositions can exclude each other only if they have the same subject; in which case they can be represented as propositional functions in the same—possibly very restricted—universe of discourse. Hence it is appropriate that the restricted Disjunction Rule should apply only to propositional functions.

We are now in a position also to prove a completely general Disjunction Rule, in which we do not have to stipulate either independence or exclusiveness. The derivation is tedious, but straightforward. By De Morgan's Law

$$\text{Prob}[F(g) \lor H(g)] = \text{Prob}[\sim(\sim F(g) \& \sim H(g))]$$
$$= 1 - \text{Prob}[\bar{F}(g) \& \bar{H}(g)]$$

$$= 1 - \text{Prob}[\bar{F}(g)] \times \text{Prob}[\bar{H}(\bar{f}g)]$$
$$\text{by the Conjunction Rule.}\dagger$$
$$= 1 - \text{Prob}[\bar{F}(g)] \times [1 - \text{Prob}[H(\bar{f}g)]]$$
$$= 1 - \text{Prob}[\bar{F}(g)] + \text{Prob}[\bar{F}(g)] \times \text{Prob}[H(\bar{f}g)]$$
$$= \text{Prob}[F(g)] + \text{Prob}[\bar{F}H(g)]$$
$$= \text{Prob}[F(g)] + \text{Prob}[H(g)] \times \text{Prob}[\bar{F}(hg)]$$
$$= \text{Prob}[F(g)] + \text{Prob}[H(g)] \times [1 - \text{Prob}[F(hg)]]$$
$$= \text{Prob}[F(g)] + \text{Prob}[H(g)]$$
$$- \text{Prob}[H(g)] \times [\text{Prob}[F(hf)]$$

and so

$$\text{Prob}[F(g) \lor H(g)] = \text{Prob}[F(g)] + \text{Prob}[H(g)] - \text{Prob}[FH(g)].$$

which is the general form of the Disjunction Rule. We can also express it, more symmetrically but less usefully, as

$$\text{Prob}[F(g) \lor H(g)] + \text{Prob}[F(g) \& H(g)]$$
$$= \text{Prob}[F(g)] + \text{Prob}[H(g)].$$

The restricted Disjunction Rules follow immediately: for if $F(g)$ and $H(g)$ are independent,

$$\text{Prob}[FH(g)] = \text{Prob}[F(g)] \times \text{Prob}[H(g)],$$

and we get the form we proved at the end of Chapter III;‡ and if $F(g)$ and $H(g)$ are exclusive $\text{Prob}[FH(g)] = 0$. From these rules we could go on to formulate probability rules for all the logical constants of the propositional calculus, and thus complete the logical syntax which probability statements must have if they always assign numerical values to properties or propositional functions themselves subject to Boolean operations.

Let us now review the argument of the last two chapters. We have justified a set of axioms sufficient for the traditional calculus of probabilities as being the reasonable, and sometimes the only, ones we can adopt if probabilities are regarded as a continuous interpolation between the extremes of truth and falsehood. The only conventions we have adopted are that true propositions should have a greater probability-value than false ones, and that the Conjunction Rule should be written in its canonical form

† The symbol $\bar{f}g$ means a-g-that-is-not-an-f; see above p. 59.
‡ p. 45.

rather than in some more complicated variant of it. The assumptions are "qualitative" rather than quantitative: they are topological ones, like Continuity, rather than specific postulates. Most of them are highly acceptable, and even the assumptions of differentiability can be argued for, and if still unacceptable, often dispensed with. We can claim that our method, τὰς ὑποθέσεις ἀναιροῦσα,† has established the traditional calculus not merely by *fiat*, nor simply as an uninterpreted portion of additive set-theory, but as *the* way to calculate probabilities, securely founded on our basic principles of arithmetic and logic. If we are to have a language of probabilities at all, we have shown what its syntax must be, in order that it may be internally consistent. But we have yet to develop the semantics of probability and give rules for assigning them in actual cases. For this we need Bernoulli's Theorem.

† Plato, *Republic* VII, 533c8. "Eliminating axiomatic assumptions".

V

BERNOULLI'S THEOREM†

ONCE we have rules for conjunction, negation, and exclusive disjunction, we can calculate the probabilities of complex propositions. In particular, we can work out important theorems where we have a large number of independent propositions which are all instances of the same propositional function, and all have the same probability: for example, that *this* toss of the coin will come down heads, and that the *next* toss will, and that the *umpteenth* toss will. A set of propositions satisfying the two conditions of independence and similarity is known as a Bernoulli sequence, or Bernoulli trial. The immense importance of Bernoulli sequences stems from the fact that in them the proportion (or frequency, as it is usually called) of instances actually possessing a feature is likely, if the sequence is long enough, to approximate to the probability of the corresponding propositional function. Bernoulli's theorem thus provides some sort of connexion between the probability of a propositional function and actual statistical *data*, and I shall endeavour to make the argument go in the reverse direction, from statistical *data* to the assignment of probabilities. But Bernoulli's theorem is easily misunderstood, and it is a matter of great dispute whether it will support any argument from statistical *data* to probabilities. In order not to misunderstand Bernoulli's theorem, it is necessary to understand how it may be proved. I follow Kneale in giving a mathematically simple, although tediously longwinded, proof of the theorem. The mathematical reader can safely skip it, and resume the philosophical argument on p. 83.

If we have a coin, and the probability of its coming down heads

---

† James Bernoulli proved several theorems. The one we here discuss is his Central Limit Theorem. It is also known as the Law of Large Numbers. There are, in fact, a whole series of Laws of Large Numbers. Bernoulli's is the weakest and mathematically the simplest; but adequate for our philosophical purposes.

this toss is $\frac{1}{2}$,† and similarly for the next one, and every other one, then it is clear that if we toss it twice there are four possibilities, namely HH, HT, TH, and TT, where HT represents the first toss coming down heads and the second tails, and so on. In this particular case we can calculate the probability of each combination by means of the Conjunction and Negation Rules: since the probability of the first toss coming down heads is $\frac{1}{2}$ and that of the second toss coming down heads is $\frac{1}{2}$, and they are independent, it follows by the restricted Conjunction Rule that the probability of their both coming down heads is $\frac{1}{2} \times \frac{1}{2} = \frac{1}{4}$. Moreover, since a coin comes down tails if and only if it does not come down heads, it follows that the probability of the coin's coming down tails is $1 - \frac{1}{2}$. Hence the probability of the first toss being heads and the second tails—HT—is $\frac{1}{2} \times (1 - \frac{1}{2}) = \frac{1}{4}$. Similarly the probability of TH is $(1 - \frac{1}{2}) \times \frac{1}{2} = \frac{1}{4}$, and TT is $(1 - \frac{1}{2}) \times (1 - \frac{1}{2}) = \frac{1}{4}$. If we are interested, not in the order, but only in the number of heads and tails, we combine the cases HT and TH, and say that there is, by the Disjunction Rule, $\frac{1}{4} + \frac{1}{4} = \frac{1}{2}$ probability of having one of each. A similar line of reasoning shows that for three tosses, the probabilities are $\frac{1}{8}$ for three heads, $\frac{3}{8}$ for two heads and one tail (in any order), $\frac{3}{8}$ for one head and two tails (in any order), $\frac{1}{8}$ for three tails.

It is easy to generalise. If all the propositions have not $\frac{1}{2}$ but α as their probabilities, their negations will have not $\frac{1}{2}$ but $1 - \alpha$ as theirs, and the respective probabilities for two tosses will be

$$\alpha^2 \qquad 2\alpha(1 - \alpha) \qquad (1 - \alpha)^2$$

and for three

$$\alpha^3 \qquad 3\alpha^2(1 - \alpha) \qquad 3\alpha(1 - \alpha)^2 \qquad (1 - \alpha)^3$$

Pascal gave a simple way of calculating the coefficients. We construct a triangular array of whole numbers as on the next page. The rule for writing each new row is that the numbers are to be written not immediately below those in the row above but under the gaps between adjacent numbers; and each number is to be the sum of the two numbers diagonally above it in the previous row.

† Strictly speaking, we should talk of the probability *of the proposition that* this coin will come down heads this toss. But it is stylistically easier to use the verbal noun *of its coming down heads* (or, more briefly, *of heads*) and I shall continue to do so.

```
                    1

               1         1

          1         2         1

     1         3         3         1

1         4         6         4         1

  •     •       •       •       •       •
```

This sum gives the number of different ways we can reach a given end result. Thus in the bottom row above we could reach the result of three heads and one tail *either* because we already had had three heads and now had one tail, *or* because we had had two heads and one tail and now had another head; but there were three ways in which two heads and one tail could happen, and so now there are four ways in all in which we could get three heads and one tail. The different routes to that particular end result are

The numbers given by Pascal's triangle are known as Binomial Coefficients. We use the symbol $^{n}C_{r}$ to denote the rth from the left in the $(n + 1)$th row from the top.†

$$^{n}C_{r} = \frac{n!}{r!(n - r)!},$$

for there are $n!$ ways of arranging n things in different orders (n ways of filling the first place, $n - 1$ the second, since one thing is

† $n + 1$, not n, because we have written down as our top row the case corresponding to 0 tosses. We could have omitted it and spoken then of the nth row, but it would have made Pascal's triangle less of a triangle and more difficult to explain.

already bespoken for the first, $n - 2$ the third, *etc.*), but since we are not concerned with orders we count as the same all the arrangements in which the r things are arranged among themselves in a different order—and therefore divide by $r!$—and similarly all the arrangements in which the remaining $n - r$ things are arranged among themselves—and therefore divide by $(n - r)!$. Dividing numerator and denominator by $(n - r)!$, we obtain

$$^{n}C_{r} = \frac{n(n - 1)(n - 2)\ldots(n - r + 1)}{r(r - 1)(r - 2)\ldots3.2.1}.$$

Alternatively we can use Pascal's rule and mathematical induction. For by Pascal's rule, $^{n+1}C_{r} = {}^{n}C_{r-1} + {}^{n}C_{r}$. If the formula for $^{n}C_{r}$ holds for n

$$^{n+1}C_{r} = \frac{n(n - 1)\ldots(n - r + 2)}{(r - 1)\ldots3.2.1} + \frac{n(n - 1)\ldots(n - r + 1)}{r(r - 1)\ldots3.2.1}$$

$$= \frac{n(n - 1)\ldots(n - r + 2)}{r(r - 1)\ldots3.2.1}[r + n - r + 1]$$

$$= \frac{(n + 1)n(n - 1)\ldots(n + 1 - r + 1)}{r(r - 1)\ldots3.2.1}.$$

Thus, if the formula works for n, it works for $n + 1$; and it does work for 1; so it works for every n.

If we look at Pascal's triangle we notice that the numbers in the middle increase very rapidly as we descend to lower rows. As n gets larger the Binomial Coefficients begin to look like a mountain. Figure 2 is constructed from the thirteen terms of the thir-

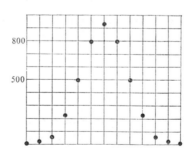

FIG. 2.

teenth row of Pascal's triangle, which work out as 1, 12, 66, 220, 495, 792, 924, 792, 495, 220, 66, 12, 1. We need to be very careful, however: intuitions are not always reliable, and in this particular case peculiarly susceptible to misinterpretation. Although the graph of the Binomial Coefficients does indeed approximate to the figure just given, the degree of apparent concentration depends on the scale of the X-axis (the horizontal axis). For as n increases, the number of possibilities spreads out wider and wider. It is NOT true that an increasing proportion of the total for each row is concentrated in the middle terms, IF we understand by 'middle terms' the one (or the two) terms occupying the exact middle of the row. For $n = 2$, the Binomial Coefficients are 1 2 1 respectively, the middle term contributing half of the total: for $n = 4$, the Binomial Coefficients are 1 4 6 4 1, and the middle term contributes only $\frac{3}{8}$ of the total. What is true, and is the key to Bernoulli's Theorem, is that if we take the middle terms not as a certain *fixed number* of terms but as *a proportion* of the total number of terms, there *is* a concentration at the middle.

It is a question of scale. In figure 2, we have the absolute value of the Binomial Coefficients, each square on the Y-axis representing 100, each square on the X-axis representing a single separate term. If we want to give not the absolute value of each Binomial Coefficient but its relative contribution to the whole, we shall have to divide each absolute value by 4,096 which is the total of them all, and this would flatten out the curve. Figure 3 shows

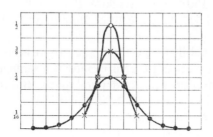

FIG. 3.

the third, fifth, and thirteenth rows on this scale—each square on the Y-axis (the vertical axis) representing $\frac{1}{16}$ of the total, the X-axis as before. If, however, we give the X-axis the same treatment as the Y-axis, the position is restored, and as we increase n, although the individual contribution of each term, even the middle one, to the total is reduced, the contribution of those which, on the new scale, lie near the middle is increased. Figure 4 gives an impression of what happens, BUT should be treated with great caution. In the figure the area under the successive curves is the same, and is more and more concentrated round the middle

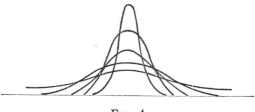

Fig. 4.

—the "mountain"—and if we specify any width of band around the middle, we can find an n for which the corresponding curve has all its "peak", comprising nearly all its area, inside the specified width, and hardly any in the "plain". But although figure 4 gives a substantially true picture of what is happening, it cannot be relied upon in argument, because the Binomial Coefficients are essentially a discrete series, whereas figure 4 has to portray them as a continuum. We need to argue the case very carefully on its own merits: nevertheless figure 4 may help the reader to grasp what is going on. But before attacking Bernoulli's Theorem itself, we need to generalise the concept of Binomial Coefficient to take account of probabilities other than $\frac{1}{2}$.

Unless the probability of each of the propositions is $\frac{1}{2}$, so that the probability of its being true is exactly the same as of its being false, the concentration will not depend on the Binomial Coefficients alone and will not be exactly in the middle. In such cases it will depend also on the probability of the propositional function in question. Let us suppose that the probability of this propositional function is α; then every diagonal step down to the left in Pascal's triangle (p. 75) will have a probability of α, and every step down to the right will have a probability of $1 - \alpha$, and so the probability of there being out of a total of n cases exactly r in which the proposition comes true will be

$$^nC_r\alpha^r(1 - \alpha)^{n-r} = \frac{n!}{r!(n - r)!}\cdot\alpha^n(1 - \alpha)^{n-r}.$$

Let us abbreviate this as ${}^{n}T_{r}$. Then the ratio

$$\frac{{}^{n}T_{r}}{{}^{n}T_{r-1}} = \frac{n!\,\alpha^{r}(1-\alpha)^{n-r}}{r!(n-r)!} \cdot \frac{(r-1)!(n-r+1)!}{n!\,\alpha^{r-1}(1-\alpha)^{n-r+1}}$$

$$= \frac{(n-r+1)\alpha}{r(1-\alpha)}.$$

This ratio decreases as r increases. It will change from having been greater than or equal to 1 to being less than or equal to 1 when

$$\frac{(n-r+1)\alpha}{r(1-\alpha)} \geqslant 1 \quad \text{and} \quad \frac{(n-r)\alpha}{(r+1)(1-\alpha)} \leqslant 1$$

that is, when r (which must be a whole number) is greater than or equal to $(n+1)\alpha - 1$ and less than or equal to $(n+1)\alpha$, that is to say, when r is the largest whole number less than or equal to $(n+1)\alpha$. We call this whole number l, and the largest term ${}^{n}T_{l}$.[†]

The ${}^{n}T_{r}$'s replace the ${}^{n}C_{r}$'s, but the picture given by figure 4 holds good, except that the peak in the middle is displaced from $\frac{1}{2}$ to about α.[‡] Our strategy[§] will be to adopt the scale of figure 4, which is essentially independent of n, and is in terms of proportions. Along the X-axis we define the middle in terms of a certain

[†] If $(n+1)$ is a whole number, so is $(n+1)\alpha - 1$, and there are two terms, ${}^{n}T_{r-1}$ and ${}^{n}T_{r}$, that are larger than any others: otherwise, there is only one whole number between $(n+1)\alpha - 1$ and $(n+1)\alpha$. For a fuller exposition, see W. C. Kneale, *Probability and Induction*, Oxford, 1949, §28, pp. 134–5, to which I am greatly indebted.

[‡] On this scale the peak will be at l/n. But l is the whole number between $(n+1)\alpha - 1$ and $(n+1)\alpha$; so l/n lies between $\alpha - (1-\alpha)/n$ and $\alpha + \alpha/n$.

[§] The standard proof is much shorter, but depends on Stirling's formula; see *e.g.* W. Feller, *An Introduction to Probability Theory and its Applications*, Vol. I, New York, 1957, pp. 50–5, 169; for a discussion of Stirling's Formula, see P. A. P. Moran, *An Introduction to Probability Theory*, Oxford, 1968, §1.12, pp. 19–23. The proof here given is adapted from W. C. Kneale's version of Bernoulli's original proof. See *Ars Conjectandi*, Pt. IV, Ch. V and *Probability and Induction*, Oxford, 1949, §29, pp. 136–8. As Kneale observes, for a book of this sort simplicity is more important than brevity. A short and very elegant proof can be based on Tchebychev's Inequality, but involves the concepts of Expectation and Standard Deviation. It is given, in a tediously expanded but conceptually simplified form, in Appendix I.

antecedently specified band around α (shaded in figure 5); and we then show that the proportion of the total contributed by the terms outside the middle, the way we have defined it, depends on *n*, and decreases without limit as *n* increases. The tactics will be to fight on each slope where the champion of the mountain can take on all comers from the plain, and get the measure of all of them together. We then let down the champion himself, and use the argument, already implicit in figure 3 (on p. 77), that any individual term, however much superior to all the lesser breeds without the pale, must itself diminish indefinitely as the total number of terms increases. The remaining terms, all on the mountain, are left in possession of the field.

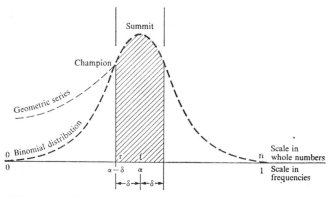

FIG. 5. The battle of the champion: *n* is taken as fixed for the duration of this battle.

The battle begins with a challenge. I challenge you, the reader, or any one else who doubts Bernoulli's Theorem, to name his standards of middle-ness and of totality, which I am to achieve, being allowed to make my *n* as large as I please. You are to name any positive number, δ, however small—provided it is positive and not zero—which specifies how close the frequency is to be to α, or, in the figure, how narrow the middle is to be. You are also to name any positive number, ε, however small—but again provided it is positive and not zero—which specifies *how close to 1* the probability is to be of some frequency within the specified limits actually occurring, or in the figure, how large a proportion of the total area beneath the curve is to be in the middle. When you have picked up my glove and specified your δ and ε, I choose two champions, one on each side, who will defend the middle-ground, taking advantage of the slope. Let us consider the champion who is defending the left-hand side of the mountain. Since we are going to be concerned not with actual numbers but with proportions, we identify the champion in terms of distance, δ, away from the summit, at approximately α, on the scale of figure 4 (on p. 78); that is, we select an arbitrary real number δ greater than 0 and less than α, and let the term ^nT_r, where r is the greatest whole number less than $(n + 1)(\alpha - \delta)$, be the champion. Then, as before (but turned upside down),

$$\frac{^nT_{r-1}}{^nT_r} = \frac{r(1 - \alpha)}{(n + 1 - r)\alpha}.$$

Since $r < (n + 1)(\alpha - \delta)$, the numerator of this fraction is less than $(n + 1)(\alpha - \delta)(1 - \alpha)$ and the denominator is greater than $(n + 1)(1 - \alpha + \delta)\alpha$, and so the whole fraction is less than

$$\frac{(\alpha - \delta)(1 - \alpha)}{(1 - \alpha + \delta)\alpha}$$

which we may abbreviate as k, noting that since the numerator is δ less than the denominator, $k < 1$. Hence we have

$$^nT_{r-1} < k \times {}^nT_r.$$

We have already seen (on page 79) that

$$\frac{^nT_r}{^nT_{r-1}} < \frac{^nT_{r-1}}{^nT_{r-2}},$$

i.e.

$$\frac{^nT_{r-2}}{^nT_{r-1}} < \frac{^nT_{r-1}}{^nT_r}$$

which is itself less than k;

$$\therefore \ {}^nT_{r-2} < k \times {}^nT_{r-1}$$
$$\therefore \ {}^nT_{r-2} < k^2 \times {}^nT_r.$$

It is clear that this argument can be generalised, in view of what we have already proved about the ratio of each term to its neighbours. In figure 5 we show the fight of the champion with all comers. The heavy dotted line shows the Binomial Distribution. The light line shows the series $k \times {}^nT_r$, $k^2 \times {}^nT_r$, *etc.* The light dotted line dominates the corresponding part of the heavy dotted line, since $^nT_{r-u} < k^u \times {}^nT_r$. But we can easily sum the terms of the light dotted line. They form a Geometrical Series,

$$k \times {}^nT_r + k^2 \times {}^nT_r + \ldots + k^u \times {}^nT_r$$

whose sum is

$$\frac{1 - k^{u+1}}{1 - k} \times {}^nT_r$$

which, irrespective of u, is less than

$$\frac{1}{1 - k} \times {}^nT_r.$$

Since $k < 1$, and depends only on α and δ and not on n, $1/(1 - k)$ is finite and bounded. Hence ^nT_r, the champion, has the measure of the light dotted line, and so *a fortiori* of the heavy dotted line. More formally, we have shown that

$$\sum_{u=1}^{r} {}^nT_{r-u} < \frac{1}{1 - k} \times {}^nT_r.$$

Of course $1/(1 - k)$ may be very large, and the sum of all the Binomial Terms may be much larger than ^nT_r itself. But it will always be bounded by $[1/(1 - k)] \times {}^nT_r$; and as n increases, $1/(1 - k)$ will not be affected, while ^nT_r will itself *decrease*. For as n increases, there will be more and more whole numbers between $(n + 1)\delta$ and $(n + 1)\alpha - 1$, and therefore more and more terms between the champion and the summit, each of which is greater than the champion. Since the sum total of all the terms together is always 1, it follows that ^nT_r must tend to zero as n increases, after the manner of figure 3 on p. 77. And therefore $[1/(1 - k)] \times {}^nT_r$ must tend to zero too, since $1/(1 - k)$ is independent of n: and so also

$$\sum_{u=1}^{r} {}^nT_{r-u}.$$

An exactly similar argument applies on the right-hand side. Hence the total of all the terms in the plain—what statisticians often call the "tails" of the distribution—on both the left- and the right-hand sides tends to zero as n increases; that is, for any specified ϵ greater than zero, the total is less than ϵ for sufficiently large n; and, correspondingly, the total of terms in the mountain is greater than $1 - \epsilon$. In the symbolism of formal logic,

$$(\delta)(\epsilon)(\exists N)(n): n \geqslant N \;.\supset.\; S(\delta, n) > 1 - \epsilon$$

where we write $S(\delta, n)$ for the sum of all the terms of the mountain, that is

$$\sum_{u \geqslant (n+1)(\alpha-\delta)}^{u \leqslant (n+1)(\alpha+\delta)} {}^{n}T_{u}.$$

If you specify what is to count as the mountain—by choosing δ—and how close to 1 the total of the mountain terms is to be—by choosing ϵ—I can find a large number—N—beyond which the Binomial Distribution will satisfy your requirements. Or, translating back into the language of probabilities, if you specify how close to α the frequency must be to be acceptable, and how close to 1 you want the probability of our having an acceptable frequency to be, I can find an N, so that in any Bernoulli sequence with N or more instances, the probability of the frequency being an acceptable one is within the degree of closeness to 1 that you have desired. This concludes the proof of Bernoulli's Theorem.

Bernoulli's Theorem is of central importance in the theory of probability. It provides the link between probability-judgements and the rest of discourse, and secures that the probability calculus should not be merely an uninterpreted calculus, but should be tied in with the rest of discourse; as Bernoulli himself said, "new it is, and difficult; but of such excellent use, that it gives a high value and dignity to every other Branch of this Doctrine".[†] For, in De Moivre's words:

> From this it follows, that if after taking a great number of Experiments, it should be perceived that the happenings and failings have been nearly in a certain proportion, such as of 2 to 1, it may safely be concluded that the Probabilities of happening or failing at any one time assigned will be very near in that proportion, and that the greater the number of Experiments has been, so much nearer the Truth will the conjectures be that are derived from them.[‡]

[†] James Bernoulli, *Ars Conjectandi*, Part IV, Ch. 4; quoted and translated by Abraham De Moivre, *The Doctrine of Chances*, 3rd ed., London, 1756, Problem LXXIII, Remark II, p. 254.

[‡] *Op. cit.*, p. 242.

But the link is a tenuous and a tricky one. Bernoulli's Theorem is difficult to grasp, and very easy to misunderstand. Itself mathematically certain—it is a deductive consequence of the axioms of probability theory, as we have seen—it connects one probability with another, and thus provides a probable connection of the one probability with a band of corresponding frequencies. But it does not lead from probable premises to certain conclusions, nor does it lead from premises about frequencies to conclusions about probabilities. De Moivre himself was aware of the latter difficulty and gives the strategy for circumventing it. He gives not a direct calculation of probabilities, but an indirect argument, a *reductio ad* almost *absurdum*.

As upon the Supposition of a certain determinate Law according to which any Event is to happen, we demonstrate that the Ratio of Happenings will continually approach to that Law, as the Experiments or Observations are multiplied: so, *conversely*, if from numberless Observations we find the Ratio of the Events to converge to a determinate quantity, as to the Ratio of *P* to *Q*; then we conclude that this Ratio expresses the determinate Law according to which the Event is to happen.

For let the Law be expressed not by the Ratio *P:Q* but by some other, as *R:S*; then would the Ratio of Events converge to this last, not to the former: which contradicts our *Hypothesis*. And the like, or greater, Absurdity follows, if we should suppose the Event not to happen according to any Law, but in a manner altogether desultory and uncertain; for then the events would converge to no fixt Ratio at all.†

Objection may be taken to this strategy, since probability arguments cannot in general be inverted, but those objections can be met,‡ provided we can meet the more fundamental one that the conclusion of Bernoulli's theorem is itself only a probabilistic one, so that invoking it to elucidate the *concept* of probability seems to beg the question. If we already had the concept, we might agree that De Moivre's method of reasoning could be applied "if not to force the Assent of Others by strict Demonstration, at least to the Satisfaction of the Enquirer himself".§ Many philosophers will agree that a probability of 1 corresponds to truth, and a proba-

† A. De Moivre, *Doctrine of Chances*, London, 3rd ed., 1756, pp. 251–2.
‡ See below Ch. VIII, pp. 126–8.
§ A. De Moivre, *Doctrine of Chances*, London, 3rd ed., 1756, p. 254.

bility of 0 corresponds to falsehood, but this seems to be no use
for the explication of probabilities lying between these extremes,
because no application of Bernoulli's Theorem, however large the
number of cases considered, will actually *reach* the value of 1. We
can get as close as we please, but never actually to it. Thus Bernoulli's Theorem, it is felt, is no use to the philosopher. It merely
explains probability in terms of frequency *and* probability.†

This objection, although natural and intelligible, is not fatal. It
is sufficient for the interpretation of the probability calculus that
we can get as close to the extremes as we please. For in interpreting we are no longer idealising. In developing the syntax of probability, we had to consider what followed from the assumption
that exact magnitudes were being assigned to propositions, and we
found that the rules for operating with the logical constants had to
be essentially what they are if we were not to land ourselves in
inconsistency. Once we have a consistent syntax for the concept
of probability and turn our attention to its semantics, we are no
longer concerned with formulating rules precisely, but only
workably. We cannot demand, in our application rules, "a
greater accuracy than the subject matter will admit of",‡ and, as
we have already seen,§ our first assessments of probability were
estimates in which a certain degree of inexactitude was ineli-
minable, and it is characteristic of all probability-judgements
that they are to some extent approximate. For this reason, instead
of identifying truth and falsehood with the *points* 1 and 0, we
should identify them with the *neighbourhoods* of these points
when we are concerned with the semantics of probability. That is,
instead of having them represented by points which are approach-
able, but in the open interval (0, 1) unattainable, we blur the
mathematical points into something more, which are not entirely outside the open interval, so that they can be not only
approached but actually reached by probabilities within the open
interval—which is to say that if we consider a sufficiently large
number of cases we can *reach* the truth. Beyond some very in-

† See, *e.g.* W. C. Kneale, *Probability and Induction*, Oxford, 1949, p.140;
G. H. von Wright, *A Treatise on Induction and Probability*, London, 1951,
reprinted Paterson, N.J., 1960 ed., p. 291 both editions; or P. A. P. Moran,
An Introduction to Probability Theory, Oxford, 1968, pp. 14–15, 57.

‡ Aristotle, *Nicomachean Ethics*, I, 1098a26ff.

§ Ch. II, p. 11; Ch. III, pp. 23–4.

definite limit we regard the differences between true and highly probable and between false and highly improbable as negligible.

A neighbourhood is a special case of what topologists call an *open* set. In a metric space—for example, the real numbers between 0 and 1—the ϵ-neighbourhood of a point is the set of all points whose distance from the point in question is less than ϵ. Thus the ϵ-neighbourhood of the point $\frac{1}{2}$ is the open interval $(\frac{1}{2} - \epsilon, \frac{1}{2} + \epsilon)$; in the special cases of 0 and 1, the ϵ-neighbourhoods are the half-open intervals $[0, \epsilon)$ and $(1 - \epsilon, 1]$ respectively. It is important to note that when we talk of a neighbourhood, without specifying an ϵ, we are talking incompletely: a neighbourhood is not any particular open interval, but one as yet to be specified. In identifying truth and falsehood with the neighbourhoods of 1 and 0, we are NOT saying that there is some definite number—say 0·001—such that probabilities greater than 0·999 are the same as truth and probabilities less than 0·001 are the same as falsehood: rather, we are giving in compact form the schema of an argument, the details of which are yet to be fixed, as they vary from case to case and depend on the particular question in issue.

Open sets have the property that their complements are not necessarily (indeed, usually are not) themselves open sets. We have noted already that it does not do to assume in probability theory that all Boolean operations are legitimate, and that if $G(f)$ is well formed formula, $G(\bar{f})$ must be too.[†] Here, however, it is not complementation but infinite intersection that concerns us. Although the intersection of any finite number of open sets is itself an open set, the intersection of an infinite number of open sets is not necessarily an open set. For example the intersection of all the open sets $(\frac{1}{2} - \frac{1}{n}, \frac{1}{2} + \frac{1}{n})$, $n = 1, 2, 3, \ldots$ is the closed set consisting of the single point $\frac{1}{2}$, and similarly the intersection of all the half-open sets $[0, \frac{1}{n})$, $n = 1, 2, 3, \ldots$ is the closed set consisting of the single point 0. The distinction is important.[‡] It is often overlooked, because philosophers and mathematicians having once become sufficiently Platonist to admit features and sets to their discourse at all feel constrained to allow infinite operations on them too. In advanced treatments of probability

---

[†] Ch. IV, pp. 59–60.

[‡] For a different application, see J. R. Lucas, "The Lesbian Rule", *Philosophy*, XXX, 1955, pp. 212–13.

theory, in particular, the assumption is often explicitly and much too readily made that we are always dealing with a σ-field, or even a complete field, of sets. But it is an important fact, at least for the application of probability theory, that we have neither infinite time nor infinite patience, and that disputes about probabilities, if they are to be of any practical import, must be decidable under some condition of finitude.

The issue can be made most clear if we concentrate on falsehood and improbability. The apparent difficulty of Bernoulli's Theorem is that every conceivable frequency is consistent with every assignment of probability. It may be improbable, but it is not logically impossible, that an unbiassed coin, where the probability of a toss coming down heads is $\frac{1}{2}$, may nevertheless produce a run of 1,000 heads; to be exact, there is $1/2^{1,000}$ probability of this happening; we cannot absolutely rule out the possibility that such a *coincidence* should have occurred.† But that is nothing terrible. Most of our arguments are non-deductive, in which we cannot rule out absolutely the possibility of the conclusion being false. Notoriously with inductive arguments it is possible, logically possible that is, for the conclusion to be false in spite of the premisses' being true; but this logical possibility does not disturb our calculations. In much the same way, the bare possibility of an extraordinary coincidence should not disturb us unduly when considering how probabilities are to be applied. We cannot rule out such a possibility, but we can console ourselves with the reflection that what we cannot rule out in matters of probability is something we cannot rule out elsewhere either.

It is, however, too easy an answer to say bluntly "Coincidences don't happen" and leave it at that. Although the demand for a logical guarantee is improper, there remains the difficulty that coincidences do—very occasionally—happen, and that the theory of probability leads us to expect them with a non-infinitesimal

† We shall see later, Ch. VII, pp. 112–18, and Ch. VIII, pp. 128–34, 141–2, that the concept of coincidence is an elusive one. Coincidences are not simply highly improbable events. It depends on the question at issue and the way in which it is posed, whether an improbable event could conceivably be regarded as a coincidence or not. But here we are concerned only with the difficulty that a coincidence, however improbable it may be, might nevertheless just possibly occur. In this chapter we deal only with the argument from improbability, and leave until later the full elucidation of the concept.

probability. As a rule of thumb we may assume that they do not happen, but for a sophisticated theory we need to accommodate the possibility of the occasional coincidence occurring without having everything explained away as mere coincidence. This in fact is achieved by our rules for settling the applicability of the concept of coincidence in disputed cases. Consider again the case of the coin. Suppose a coin is thrown one hundred times, and the number of heads is considerably greater than round about fifty. Suppose, for example, that out of one hundred tosses we get seventy which turn up heads. We suspect that the coin is biassed. It is then pointed out to us that occasionally an unbiassed coin will yield in a run of a hundred tosses seventy which turn up heads. We agree that this is possible, but are reluctant to accept that this is what has happened in our case; we see no reason why our case should be such an exceptional one. Nevertheless the point is maintained against us. How then do we settle the dispute? Neither party is clearly at odds with the facts, though we feel that the opposition's thesis is an implausible one. Still, we concede that it is possible, and cannot be absolutely ruled out. How then do we decide the issue? The answer is simple. We look for further evidence; that is, usually, we toss the coin some more times. If, upon further tosses, the unusual proportion of heads is not maintained, we concede that the opposition was correct in their conjecture that this was just one of those exceptional cases, which we expect occasionally to occur, where the proportion of a given number is much more than one half. If, however, the proportion of heads remains at about the unusual figure reached in the first hundred, we stick to our original opinion. Another illustration will make the same point: we come across a bundle of type-written sheets, the first one of which is entitled "The Plays of Shakespeare", and the second of which has the opening lines of "Twelfth Night". We infer, reasonably enough, that this is a type-written copy of the plays of Shakespeare. It is pointed out to us, however, that a monkey, taught to type and set to type for a sufficiently long time, would sooner or later produce just such a pair of pages. How can we be sure that this is not what has happened with our bundle? Once again, this thesis, if seriously maintained, can be rebutted in the same way: we *go on*, and read the next page. If it, and subsequent ones, are all gibberish, we agree that something funny has happened, and might even be persuaded that it was as the opposition claimed. But if subsequent pages continue "Twelfth Night" and the other plays in the accustomed manner, we dismiss the opposition's suggestion as trivial and of no weight.†

It is clear then what our procedure is for settling disputes when it is being maintained that the observations we have so far obtained form a highly exceptional conjunction of events. If such a contention is put forward seriously, we go on further. The contention cannot be ruled out by monologous argument alone;‡ for, as we have been stressing, no set of observations, and therefore not the ones we have obtained, are *inconsistent* with a probability-judgement about them: but it can be ruled out by the eliciting of further facts; for it is certain that the procedure of going on and continuing to elicit more and more instances of the type of event in question will in the end carry complete conviction, and one that is not open to any further doubt.

What is the logic of this? It would at first sight seem a very irrational procedure. For all the arguments which were used by the opposition at the outset of the argument can be used by them at the end of it also. There is still only an improbability, and not a logical impossibility, of the set of observations and the probability-judgement about them both being true. We have not converted the improbability into an impossibility; indeed we cannot, by this or any other procedure. The rationale of probability arguments cannot be assimilated to that of arguments concerning universal statements. But this does not preclude their having a rationale of their own. To see what it is, and to see why the procedure we use is not as irrational as it first appears to be, we must again see the issue as a dialogue—a dispute or argument between two *persons*, not a relation of compatibility or incompatibility between two sets of *statements*. The opposition's case was that the preponderance of heads was not due to the coin's being biassed, but was a *coincidence*: and when we demurred, it was insisted—and we could not deny it—that there was a small but definite

† These alternatives do not exhaust the possibilities. It is always possible that the next page, through an unfortunate coincidence of misprints, is gibberish, but subsequent ones are intelligible again; or that the unusual proportion of heads is not at first maintained, but later is. See further below, Ch. VIII, pp. 139–40.

‡ See above, Ch. I, pp. 3, 5–6.

chance of an unbiassed coin behaving as this one had. So we accepted the possibility of their hypothesis being true and put it to the test. The preponderance of heads was maintained, contrary to what was to be expected if the hypothesis of the opposition had been true. Of course, it is true that the further run of heads *could* be another coincidence, even more of a coincidence than the initial run. There is still, and we must still concede it, a small but definite probability of an unbiassed coin producing even a double run of heads. But, and this is a big 'but', it is not the same probability. The probability of a hundred tosses of an unbiassed coin producing 40% more than the normal number of heads is small enough in all conscience: the probability of it happening a second time is far smaller. The point of the argument, however, lies not in the fact that it is *far* smaller, but only in that it is smaller. If our opponent still maintains his doubts, he will have shifted his ground. He is no longer asking us to accept a coincidence of which there is a chance of, say, 1 in 1,000, but one of which the chance is only 1 in 1,000,000. We are prepared, since he insisted so vehemently, to entertain the possibility of its being a coincidence the first time he suggested it; but not the second. We have taken him seriously once, despite the implausibility of his thesis: we went to the trouble of obtaining further observations, which did not bear him out; so we are not going to take him seriously again.

Behind this somewhat emotional account lies a logical point. In dealing with the sort of objection which the opposition is making to our probability-judgement, we demand that they fix on some definite standard of improbability, and abide by that. If it is alleged that our case is an exceptional one, which turns up only once in a thousand times, we accept that, and seek further observations which will agree or disagree with its being the one exceptional case in a thousand. If now, in order to save their thesis, the opposition have to shift their ground and say that it is the odd case in a million, then we are quite justified in refusing to take them seriously, because clearly there is no end to the process. However long we go on, since we can have only a finite number of observations in all, there will always be a finite probability that the whole set of observations is an exceptional one, and the suggestion always could be made that this was what had happened. But that suggestion, since it always can be made, cannot always be taken seriously. It is too easy an objection to carry any weight. If it is made, I face the person who makes it with the following alternative: either you are being serious, I say, and are worried lest this should be a case in a hundred, a case in a thousand, a case in a million, or some such; if so, tell us which it is, one in a hundred, one in a thousand, one in a million, or whatever, and we will go on taking observations and test our probability-judgement up to your standards of acceptance: or you are not being serious, and are merely going to go on saying, however many observations we make, that there is still a possibility of its being a coincidence; but that is merely to give notice that you are not going to be persuaded by any evidence whatsoever, and if that is your line, we may as well stop arguing at once; in fact, there is no possibility of our having a rational argument, for you cannot mean the same by the word 'probability' as I do, since you regard probability-judgements as quite unrelated to observations and evidence, and think that they are equally compatible with every possible set of observations:† whereas for me it is essential that they should fit—though the fit need not be an exact one—the facts as we know them. I have conceded that in probability the evidence does not force us to abandon our hypotheses in the simple, knock-down way that counter-examples force us to abandon universal statements. But that notwithstanding, probability-judgements are about the world: they raise expectations about how things turn out; and if it did not matter that these expectations should be consistently disappointed, probability-judgements would cease to have any relevance at all.

It is therefore quite reasonable for us to require that our opponent does not embark upon a line of argument which would end in his denying the relevance of evidence to probability-judgements altogether, by being prepared to cushion probability-judgements against contrary evidence indefinitely. But if he is not to do that, he must not go on urging indefinitely small probabilities on us, as being possible coincidences to explain away the discrepancy between the evidence and his rival thesis which he is putting forward in opposition to ours. And if indefinitely small probabilities are ruled out, then only definite, though possibly very small, probabilities are left as being legitimate. Our opponent is allowed to dismiss, as coincidences, events or combinations

† Compare Ch. I, pp. 4–5, above.

of events the probability of which is as low as any figure greater than zero he cares to mention; provided that we can take him up on his figure, and seek further observations, and if these further observations can be reconciled with his rival hypothesis only by assuming a coincidence the probability of which is even lower than the figure he named, then he will abandon his rival hypothesis and his opposition to our hypothesis, and will accept our contention as true.† I am prepared to be reasonable and allow that my probability-judgement—say, that the coin was biassed so that the probability of a head turning up was 0·7 instead of 0·5— might be wrong, and my opponent's rival hypothesis—that the coin was unbiassed, and the probability of a head turning up was 0·5, but I had chanced upon one of those exceptional runs that do occur even with a true coin in which one side does turn up disproportionately often—was correct: but he must be reasonable too, and allow that he also may be wrong, and that it may not be a coincidence that there should be so many more heads than might have been expected if the coin were unbiassed. We are prepared to treat his allegation that a coincidence may have occurred, provided it is sufficiently definite an allegation to be testable. The willingness on my part to countenance suggestions of coincidence demands a readiness on the part of the other side to specify the degree of coincidence that could have occurred. This done, the dispute becomes a decidable one: further observations are all that are required.

In short, when dealing with probabilities, we are prepared to listen to a questioner who is unsatisfied, provided he is not unsatisfiable. The doubt "it might be a coincidence" will be considered, provided that this is not a neurotic doubt, which will continue to be put forward whatever we do, that is, provided that when the doubt is put forward as a counter to our original probability judgement, it can be discovered what evidence would satisfy the questioner and set his doubt at rest. If he will tell us that, we shall attempt to satisfy him, and be open to conviction ourselves, should the further evidence fail to support our case: but if he cannot tell us anything that would satisfy him then we

---

† That is to say, our opponent must state a small positive number ϵ, such that the interval $[0, \epsilon)$ is to be identified with falsehood for the rest of *that* argument (it does not have to be the same for all arguments; see above, p. 86, and below Ch. VIII, pp. 134–40).

must separate and continue discussion no longer, for it is clearly pointless for us to try to satisfy a person who will not be satisfied.

The dialectical root of the concept of probability thus bears important fruit even when the concept has been transplanted to monologous soil. We no longer can ask a man for his reasons for making only a guarded probability-judgement, instead of an unqualified assertion: but if any doubt is raised about our regarding very highly probable statements as true or very highly improbable ones as false, we can put the doubt to the question, and dismiss it if it is merely Cartesian. Of course, we must play fair. Doubts may be non-Cartesian. It may be less implausible to explain a phenomenon as a highly improbable coincidence than in an alternative way that is even more far-fetched.† Coincidence is a dialectical concept, and what counts as a coincidence depends on the exact way the question is put, and the argument cannot proceed without there being first a pause to give the opponent opportunity of putting forward counter-proposals. It is only if he has nothing to say, and puts forward the suggestion of a coincidence without any backing, that we deploy our present argument, which we can express mathematically by saying that the concept of coincidence is to be regarded not as an improbability of a particular value, but as a sequence or ordered set of improbabilities. Our restriction upon the possibility of coincidences can then be felicitously expressed by saying that events of any non-zero probability, however small, can occur, but not a set of events the probabilities of which have zero as a "limit point". This is equivalent to saying that the possibility of coincidence can always be urged, provided the probabilities of the coincidences which are urged do not form a sequence that converges to zero. Corresponding to our earlier requirement that there should be a sticking point, on which the opposition should take its stand, and, if forced from it, be prepared to acknowledge defeat, is now the requirement that the limit, if any, of the probabilities of the alleged coincidences, should be greater than zero, so that an explanation in terms of a coincidence the probability of which was less than this limit should be acknowledged to be no explanation at all. In the one way, we are ruling out lines of argument which fall back indefinitely on more and more improbable coincidences: in the

† See further below, Ch. VIII, pp. 128–9, 141–2, 157.

other, we are ruling out sequences of probabilities which tend to zero.

For all these reasons, it seems appropriate to identify not just the points 1 and 0 but their neighbourhoods with truth and falsehood; and then Bernoulli's Theorem will constitute a link between probabilities on the one hand and the rest of discourse on the other. It will apply when we are dealing with a large number of reproducible, independent cases which are qualitatively identical and numerically distinct. In other cases, where we are dealing with singular propositions, we can only make estimates of probability, and if our estimates disagree there is no way of putting the matter to the test. But it is not an objection that we sometimes cannot check up on our estimates, in the way in which it would be an objection if our estimates were always and necessarily incorrigible. We often are dealing with large numbers of reproducible cases, and then our interpretation rules provide an effective anchor to the concept in the rest of our discourse. Which is all that we require for the concept to be meaningful.

We have now completed the programme of generalising truth and falsehood. If we want to interpolate continuously varying degrees between truth and falsehood, we must adopt rules essentially similar to the traditional ones and then will have Bernoulli's Theorem, which will link probabilities that lie between truth and falsehood with truth and falsehood at the extremes. The relationship of probability theory with the discrete logic of truth and falsehood is a double one, like that of quantum mechanics with classical mechanics.

Quantum mechanics occupies a very unusual place among physical theories: it contains classical mechanics as a limiting case, yet at the same time it requires this limiting case for its own formulation.[†]

Similarly probability theory is both a generalisation of the ordinary logic of truth and falsehood—and therefore contains the propositional calculus as a special case—and depends on the ordinary logic of truth and falsehood to give it content. The relations of probability with truth are thus somewhat double-faced, and have led many philosophers to yearn to replace them by something simpler. In particular, it has led them to the Frequency theory.

    † L. D. Landau and E. M. Lifshitz, *Quantum Mechanics*, tr. J. B. Sykes and J. S. Bell, London, 1958, p. 3.

Bernoulli's Theorem, according to our argument, links probabilities with proportions or *frequencies*, as they are commonly called. It shows that *if* the probability is α, then if we take a sufficiently large number of cases, it is almost certain (in the sense elucidated above) that the frequency will fall near α within antecedently specified limits. The argument can be reversed. But, as we shall see, there are many pitfalls. Many philosophers have despaired of being able to use Bernoulli's Theorem, while remaining sure that we do in practice determine probabilities on the basis of observed frequencies; and therefore, as often in philosophy, seek to secure by definition what they have failed to achieve by argument. They *define* probability in terms of frequencies, and say that the probability of an event's having a certain characteristic is the limiting value of the relative frequency of events having that characteristic in a class, satisfying certain conditions,† of events of that type.‡

The Frequency theory can be regarded as an *extensional interpretation* of the theory of propositional functions given in Chapter IV. Although I have argued that we can ascribe probabilities to propositions and apply some of the rules of the calculus of probabilities, we can ascribe reasonably precise probability-values and apply the whole of the calculus of probabilities only to propositional functions. The Frequency theory takes account of this, but instead of interpreting propositional functions intensionally, defined by the meaning of the terms employed, interprets them extensionally, as the class of entities to which the terms can be correctly applied. Instead of the individual variable, 'Englishman' or 'Englishman-with-a-brown-left-eye', the Frequency theorists consider the class, Englishmen or Englishmen-with-brown-left-eyes. As an aid to understanding, it is often a helpful move, and one I shall on occasion make myself. But it is not always so; quantum mechanics is less difficult to understand

† See below, p. 98.

‡ Some writers say that Aristotle was the first to espouse the Frequency theory (see *Prior Analytics*, II, 27, 70a3, *Art of Rhetoric*, I, 2, 1357a34). J. Venn, *The Logic of Chance*, 3rd ed., London, 1888, was the first proponent in modern times. The best modern source is Richard von Mises, *Probability, Statistics and Truth*, 2nd English ed., London, 1957. See also, H. Reichenbach, *Experience and Prediction*, Chicago, 1938, Ch. V, and K. R. Popper, *The Logic of Scientific Discovery*, London, 1959, Ch. VIII and Appendices *ii-*vii.

if we banish the word *ensemble* from our vocabulary. More generally, the Frequency theory suffers from a radical incoherence of aim which disqualifies it from providing a definitive elucidation of the concept of probability. In part it aims to be a simple, commonsensical theory based on the actual practice of ordinary men: but in the hands of the mathematicians it has become a highly sophisticated theory seeking to satisfy Russell's thesis of extensionality, and attempting to construct a Platonist definition more suited to Cantor's paradise than our actual sublunary world of affairs.

The Frequency theory starts off as a simple man's theory. Its great merit is that it squares with our unsophisticated intuitions about what happens "in the long run", and our beliefs about the guidance implicit in a man's saying "in three cases out of four", as well as with our sophisticated statistical practice. But there is a confusion between the criteria for the use of a word and its meaning. The Frequency theorists are quite right in the account they give of the typical circumstances in which we make either amateur or professional statistical prophecies, but wrong in supposing that they constitute the meaning of the word. It is the same mistake as other Logical Positivists make when they analyse the meaning of the statement that Paul is in pain as an account of Paul's drawn face, clenched teeth, white knuckles *etc.* These are good grounds for believing Paul is in pain, but when on the strength of these we conclude that Paul is in pain, we are believing something further about Paul, not merely that he has a drawn face, clenched teeth, white knuckles *etc.*—else, how should we feel sympathy for him? Similarly with probability. The evidence, necessarily in the past, on which we base our probability-judgements is one thing: the judgement itself, usually referring to the future, is another.

The Frequency theory cannot deal with singular propositions. It has to re-construe them as being really about members of a class. Sometimes, of course, they are. 'Man is a rational animal', though singular in form, is plural in intent. So is 'A man aged twenty who smokes 40 cigarettes a day has a 15% probability of contracting cancer of the lung before he is fifty'. Even when the singular form is not merely an idiom but really refers to a single instance, there is still a covert universality imported into the proposition by virtue of the fact that probability-judgements must be based on reasons, and reasons are universal. If I give a $16\frac{2}{3}\%$

probability to the proposition that this particular die will come up six on its next throw, or a 15 % probability to the proposition that Smith will contract cancer of the lung before he is fifty, I must have reasons for making that assessment, and these reasons will depend on some general features of the die or of Smith's case, which therefore will define a class. True: but not enough to save the Frequency theory. For although probability-judgements, like many others, are covertly universal, it does not follow that the man who makes a probability-judgement has an antecedently specified class in mind;† nor, often, will he have culled his data from exactly that class which the instance exemplified. In assessing the probability of Smith's getting cancer, I may have to set against the fact that he smokes, the fact that he smokes a pipe, and against the fact that he lives in the country away from fumes and exhausts the fact that both his parents and his brother contracted cancer of the lung. These, and possibly further factors too, will have to be weighed against one another before I arrive at an assessment: and it may well be that there are no statistics for the relevant combination of factors, and I have simply to use my judgement. But that does not affect the meaning of what I say, although it may affect its reliability. And therefore I cannot be saying that the frequency of actual Smith-like persons contracting cancer before they are fifty is 15 %: at the most I am committed to the hypothetical claim that if there were any such, the proportion of them contracting cancer before the age of fifty would be about 15 %. But with the loss of actuality, the classes of the Frequency theorist lose their charm.

Besides, not all singular propositions can be universalised. Some are universal already, and logically unique. If I say that the Theory of Special Relativity is 99·9 % probable and the Theory of General Relativity only 70 %, I cannot be talking of the frequency with which theories of these two types respectively turn out to be true. No appeal to universalisability, no radical reconstruction of such probability-judgements is possible: the Frequency theorist cannot accommodate them, he must dismiss them altogether, as not being really judgements of probability at all. Again, he can make some show of reason. As we have seen, such judgements can

† See further, J. R. Lucas, "The Lesbian Rule," *Philosophy*, XXX, 1955, pp. 205–9.

only be estimates, and fairly rough ones at that. We cannot usefully distinguish between 99.9% and 99·85%, nor between 70%
and 66·6%. But nor need we. It is sufficient that some sense can
be given to some assessments, however crude, of the probability of some theories. Unless we take a Procrustean line of refusing to admit as probability-judgements any that do not
measure up to our Frequency theory requirements, it is difficult
to deny that at least sometimes we make probability-judgements
about logically singular propositions. And if so, the Frequency
theory—which is a theory about what probability-judgements
mean, and therefore must be applicable in every case—cannot be
correct.

There are other difficulties. Not every class has a reasonably
stable proportion of its members possessing or not possessing a
given characteristic. Von Mises formulates two conditions which
a class must satisfy before it can be what he terms "a collective",
and statistically it is sound practice to apply these two criteria
before setting out to extract any probabilistic conclusions: but
both are very awkward to incorporate into a definition. The first is
a condition of convergence: for a class to count as a collective, the
relative frequency of the members having the property in question
should tend to a fixed limit as the number of members in the class
is indefinitely increased.† The second is a condition of randomness: for a class to count as a collective, there should be no way of
antecedently selecting a sub-class in which the relative frequency
of the members having the property in question tends to some
other limit than that for the whole class. Since nothing is given
about the individual members of the class except the order in
which they are to be considered, we can express the second condition by saying that "place-selection" should leave the limit of
the relative frequency unaltered.‡ When incorporated into a definition, 'indefinitely increased' in the first condition is taken to
mean 'infinite', and sophisticated difficulties about transfinite
cardinals are urged. The concept of randomness in the second
condition is likewise open to objection: in any finite class there
are necessarily some place-selections which will alter relative

† Richard von Mises, *Probability, Statistics and Truth*, 2nd English ed.,
London, 1957, pp. 14–15.
‡ *Op. cit.*, pp. 23–8.

frequencies, and in an infinite class it is unclear what relative frequency means.

These mathematical difficulties, although serious, are surmountable. Transfinite arithmetic has some useful Boolean features.[†] The difficulty about defining a ratio (other than zero, unity, or infinity) arises only with cardinal transfinite numbers: but von Mises defines his collective as a sequence,[‡] not an un-ordered class, and the stock objections do not apply, although it remains difficult to formulate watertight definitions for concepts like the Césaro sum of an infinite series. Again, a convinced mathematical Platonist should have no difficulty in believing in the existence of random sequences; for there are $2^{\aleph_0}$ possible infinite sequences that might constitute a von Mises Collective, but only $\aleph_0$ for which the rule specifying them could be antecedently given in a finite number of terms.[§] As an exercise of pure mathematics, the formulation of a Frequency theory may still be feasible, mathematical difficulties notwithstanding. But it has ceased to be a plain man's theory. We are no longer being offered a reductive analysis of probability in terms of observed frequencies in actual classes, but a complicated mathematical analysis involving limits, infinities and other highly abstract entities. Frequency theory has become a mathematician's pastime instead of a plain man's guide.

We may further object that the Frequency theory does not altogether accord with the traditional concept of probability. Sometimes a long run, provided that it is only finitely long, should yield a frequency diverging from the probability. Only if we escalate to infinity can we apply the Strong Law of Large Numbers and say that it is "almost certain" that in an infinitely long run the frequency will converge to the probability. But for any actually observable long run, however long it may be, there is a non-zero, though small, probability of the frequency's subsequently substantially diverging from the limit it had apparently been converging to. Such possibilities are difficulties for any

† Two cardinals of the same power are idempotent under addition and multiplication, e.g. $\aleph_0 + \aleph_0 = \aleph_0$, and $\aleph_0 \times \aleph_0 = \aleph_0$. Also $\aleph_0 + 0 = \aleph_0$ and $\aleph_0 \times 0 = 0$. The analogue for complementation is less satisfactory.

‡ p. 12.

§ Alonzo Church, "On the concept of a random sequence", *Bulletin of The American Mathematical Society*, XLVI, 1940, pp. 130–5, gives a mathematically rigorous definition.

theory of probability, but fatal for the Frequency theory because it has ruled them out by definition, and definitions are hard, un-yielding and brittle. In my own account I needed to give a lengthy argument to show how I could accommodate the possibility of coincidences. I give a rationale for rejecting the possibility of an indefinitely extended coincidence, and thus a justification for actually using Bernoulli's Theorem to argue from frequencies to some conclusion about probabilities. But I do not exclude the possibility of my being mistaken. In any particular case I may be tripped up by a coincidence that actually has occurred. I can be wrong in particular cases: and therefore my definition or proba-bility need not be. But with the Frequency theory there is no give. The Frequency theorist does not argue his way from the evidence to the conclusion, but defines it; and if any incongruity emerges, it proves the definition faulty, not merely a particular judgement on a particular occasion. And definitions need to be infallible, or they are futile.

The Frequency theory is to be rejected not on mathematical, but on philosophical, grounds; and in particular because it is both otiose and on occasion wrong. It is otiose because Bernoulli's Theorem already provides a link between probabilities and fre-quencies: and it is sometimes led into error because it attempts to secure by stipulation what needs to be achieved by honest argu-ment, and stipulative stealing is to be rejected not so much on Russell's moral grounds as because it suffers from the disadvan-tage of extreme fragility.

VI

SINGULAR PROPOSITIONS

I T was a demerit of the Frequency theory that it could not deal with singular propositions. It is a merit of my account that it can. If probabilities are a continuous interpolation between truth and falsehood, since any proposition can have a truth-value, it can have a probability-value also. Nevertheless, there are difficulties. As the last two chapters have shown, it is propositional functions rather than propositions that make the running in probability theory. We can prove the restricted Conjuction Rule for independent propositions, and the Negation Rule; but if we want to prove the unrestricted Disjunction Rule—and, obviously, the unrestricted Conjunction Rule, the so-called rule of Conditional Probabilities—we need to consider propositional functions, not just propositions. And Bernoulli's Theorem clearly has no relevance to singular propositions. Therefore when we are dealing with genuinely singular propositions we have only part of the calculus of probabilities available, and can only estimate their probability-values very roughly, without the aid of any mathematical argument.

For some singular propositions this is all we need say. These are the propositions which are necessarily singular because they are about universals—theories for example. It is perfectly legitimate to talk of the Theory of Special Relativity being probable, and of its being more probable than the Theory of General Relativity: but although we can say that the latter is more probable than not, we cannot decently say that it is 54% probable or 87·5% probable. Nor can we use the unrestricted Conjunction Rule, nor the Exclusive Disjunction Rule, nor the definition of Conditional Probabilities. For they apply only to propositional functions, where it makes sense to talk of changing the universe of discourse. Very occasionally we can, as we shall see later,† construe theories as ranging in a universe of discourse: but these are

† In Ch. VIII, pp. 149–57.

all special cases where the general form of the theory is fixed, and it is only some parameters that are variable. In these special cases we can apply the unrestricted Conjunction Rule to assign probabilities in accordance with Bayes' Theorems: but even then we are assuming that the general form of the theory is fixed, and this is a singular proposition about which we might in turn have to make a probability-judgement. These special cases apart, we may want to assign probabilities to singular propositions on certain conditions. I might want to say that a zero probability should be assigned to the Theory of Special Relativity if the Michelson–Morley experiment were to yield a positive result, but a high one if a negative result. Some thinkers have extended the Conjunction Rule to such cases, and would consider it a demerit of my approach that we cannot say,

The probability of the Theory of Special Relativity is the arithmetical product of the probability of a negative result to the Michelson–Morley experiment and the probability of the Theory of Special Relativity under the condition of there being a negative result to the Michelson–Morley experiment.

But such a locution rings awkwardly in our ears. Theories are not like that. They are no good unless they connect a number of *different* phenomena. There must be more than one *route* leading to a theory, and therefore more than one argument for it; and we cannot assess its probability by considering only one argument for it, but them all, possibly including some which tell against it. We cannot "add" conditions, or hope to add up the probability that a theory obtains from the various arguments in favour; and in any case the weight of argument is more than cumulative, and we cannot estimate exactly, or often even at all, the contribution of different considerations separately. The Special Theory of Relativity obtains support not simply from the Michelson–Morley experiment, but because it unifies electromagnetic theory in a coherent way and seems to be a natural generalisation of the principle of Galilean relativity which we have already in classical mechanics. We cannot isolate the contribution from the Michelson–Morley experiment to the creditworthiness of the Special Theory. To assess the probability of a theory, in the face of diverse arguments for, and possibly against, demands judgement, and sometimes genius, not the calculus of probabilities.

If a further argument is needed to show that the Rule of Conditional Probabilities does not apply, we can note that if it did, it would not matter which *route* we used to reach a conjunction of theories. To borrow Jeffreys' notation, $P(h/e)$ for the probability of hypothesis h on evidence e,† suppose we have two hypotheses h and g, where g is a grander one for which h is evidence, but there are other arguments for g and g could be true even if h were false; then by the Rule of Conditional Probabilities,

$$P(h \ \& \ g/e) = P(h/e) \times P(g/h \ \& \ e)$$
but also $\qquad\qquad = P(g/e) \times P(h/g \ \& \ e).$

In this second line, e is evidence for g only because it is evidence for h. $P(g/e)$ therefore is the same as $P(h \ \& \ g/e)$. But $P(h/g \ \& \ e)$ is not 1, because h could be false even though g and e are true. The suffocation of mice in closed vessels was evidence for the phlogiston hypothesis, and the phlogiston hypothesis was evidence for the theory that the energy of chemical reactions can be dissipated as heat, but granted the latter theory and the suffocation of mice the probability of the phlogiston theory is far from 1. Or, to take another example, where the asymmetry goes the other way, we can say that the observations of Tycho Brahe were evidence for Kepler's three laws of planetary motion, and that Kepler's laws were themselves evidence for Newton's three laws of motion and law of gravitation, although not perfect since Kepler's laws could be true even if Newton's were not. (Although if Newton's laws are true, Kepler's are a deductive consequence.) Tycho Brahe's observations are thus evidence for Newton's laws as well, but the probability of Newton's laws on their evidence should be less than that of Kepler's laws, if the rule of Conditional Probabilities is to apply. But in fact Newton's laws seem no less probable than Kepler's, not because there is other, non-astronomical, evidence in favour of them—in Newton's time there was very little—but because they explain Kepler's laws in a systematic and coherent way. Kepler's and Newton's laws together have a greater probability than Kepler's laws by themselves, and the rule of Conditional Probability cannot possibly apply.

It is with singular propositions which are not necessarily singular, however, that confusion arises. For these contain referring

† Harold Jeffreys, *Theory of Probability*, 3rd ed., Oxford, 1961, p. 20.
8—c.o.p.

terms, which refer adequately without describing fully. We can refer to Jeffreys' friend Smith,† or to tomorrow, or to that man there without knowing all the relevant facts about Smith, or to-morrow, or that man there. And hence it is easy to change the description under which we are considering Smith, or tomorrow, or that man there, without realising it, in a way which cannot happen when we are talking about universals where what we are referring to is constituted by the description under which we are describing it. Barring a few inessential qualifications, if I re-describe the Theory of Special Relativity, I am talking about something different, a new theory, that theory which I have described. But if I re-describe Smith or that man there or to-morrow, I still think, rightly, that I am talking about the same person or day, only that I know more about him or it. And this is quite all right for ordinary purposes, but is dangerous when I am talking about probabilities because probabilities are assigned to propositions, not people or days or events, and a proposition about a person is changed if the description of the person is changed, even though the person himself is not. The proposition "Smith, who is an Englishman, has a blue right eye" is a dif-ferent proposition from "Smith, who is an Englishman, and who has a brown left eye, has a blue right eye". In the cases, therefore, where we really do know Smith, and know lots of things about him, it is easy to forget that the proposition whose probability we are assessing is a singular but very complicated one. If a doctor is consulted by a patient whom he knows well about his expectation of life, he will take into account many different fac-tors—his smoking, his non-drinking, his athleticism, his non-adiposity, his good heredity, *etc.*, *etc.*, *etc.* In the end he will have to hazard a guess, because he will almost certainly know more relevant factors about the man than are used to define any class for which statistics have been collected. Every individual, if we really know him, is untypical. And for people who are not types, there are no exactly applicable descriptions or exactly relevant statistics. We can only hazard estimates, not subsume them exactly into pigeon-holes.

Even so, there is a difficulty. As we acquire more relevant in-formation we shall apply the propositional function as ranging in a more fully specified universe of discourse, and as we change the

† See above, Ch. IV, p. 50.

specification of the universe of discourse, the probability of the propositional function, and therefore of the singular proposition considered as an instance of the propositional function, will change, and worse, will change discontinuously. What is the probability of a pair of true dice coming down double six?—$\frac{1}{36}$. What is the probability of its coming down double six given that the first die came down six?—$\frac{1}{6}$. What is the probability of its coming down double six given that the first die came down five?—0. What is the probability of these dice on this occasion, on which both come down a six, coming down double six?—1. The last is no different from the others, except that the universe of discourse has collapsed into a single instance, and the probability-value has collapsed into a discrete truth-value, in this example, True. We start with $A(g)$. More information leads us to specify the question more closely, $A(fg)$, $A(efg)$ ... $A(bcdefg)$, where, as before, fg is a complex name for a *single* individual variable ranging over a more specific universe of discourse, of all the g's that are also f's, that is, of all the g's that are F. As the universe is more and more specified, we may reach the stage where either the specification alone rules out the possibility of any individual in it being A—e.g., if we specify that the first die came down five we exclude a double six— or the specification rules out the possibility of any individual in it not being A—e.g., if we know that each of the dice came down a six, then a double six it is. Whether or not we reach such a stage while we are still dealing with propositional functions, we must reach it when the universe has narrowed down to a single individual and we are dealing with a proposition. For a single individual is completely specified in principle. The Law of the Excluded Middle holds: either it is A or it is not A; either we are considering $A(abc \ldots g)$, which is true, or $A(\bar{a}bc \ldots g)$, which is false. When we are no longer talking about *a* die on *an* occasion, but *this* die on *this* occasion, then the proposition is indeed a proposition, and must be either true or false.

The difficulty is most acute when we assign a probability to a singular proposition predicting a fairly ordinary future event, *e.g.*, that it will rain tomorrow; we find the probability of the proposition dancing about, and finally, when the time comes, either turning out unqualifiedly true or turning out unqualifiedly false. Whichever way we look at probabilities, we feel uncomfortable.

If we consider probabilities as analogous to truth-values, we feel that a singular proposition must, by the Law of the Excluded Middle, be either true or false, so that to assign any other *soi-disant* "truth"-value must be wrong. If, on the other hand, we reify probabilities and think of them as existing objectively in the external world, then the reified probability defies the principle of continuity, as it jumps from one value to another, and finally collapses into either truth or falsehood when the moment comes for the prediction to be either verified or falsified.

The key to the problem is the specification of the proposition. Although the same form of words is used, "It will rain tomorrow", "The next throw of these dice will be a double six", we may come to know more and more about the events referred to; and therefore propositions about them, and in particular, propositions predicting them, will be set in different universes of discourse. My meteorological description of tomorrow, Wednesday, is much less complete now at 6 p.m. on Tuesday than it will be at midnight; and in the course of tomorrow it will become more and more fully specified, until finally I can refer to it under a description which will answer in itself any question I might raise of whether it was raining on Wednesday or not. So too, when I consider the probability of the next throw of these dice being a double six, I consider it under the description of being just *a* throw of true dice: but after the event I can describe it more fully, and say whether it yielded a double six or not, and therefore I naturally consider it under that fuller description, in which it is already given whether it was a double six or not. A Laplacian calculator who knew the initial conditions of our throw, would specify the universe of discourse which included these details, and in such a universe he would be able either to assign to the propositional function "yielding a double six" the truth-value True or the truth-value False. The probabilities of singular propositions appear to change, because they are really assigned not to propositions, but to propositional functions, and the propositional functions are set in successively different universes of discourse. As we specify the universe of discourse more and more fully, it becomes narrower and narrower, with the probability to be assigned to the propositional function jumping at each narrowing, until finally the specification of the universe of discourse specifies also the truth-value of the propositional function, and we get a "collapse" of probability-values into discrete truth-values.

We may borrow a term of Professor Quine's,† and say that probability-judgements are "referentially opaque" in much the way that belief-statements are. Just as I can believe that Venus is the Evening Star without also believing that the Morning Star is the Evening Star, so I can ascribe one probability to a proposition referring to a particular under one description while ascribing a different probability to the same proposition referring to the same particular under a different description. 'It will rain tomorrow' or 'The next throw of these dice' refer to the same event as, later, 'It will rain today' or 'the last throw (which came down a double six) of these dice' do; but the different description of the day or the throw makes a different propositional function applicable, and so a different probability-value assignable. Nor is it only a verbal change that leads us to regard a different propositional function as applicable: a change of context may be enough. 'Tomorrow' said on Tuesday and 'today' said on Wednesday, and 'tomorrow' said at 9 a.m. on Tuesday and 'tomorrow' said at 11.50 p.m. all refer to the same stretch of time, but in different ways: and in their different contexts, different propositional functions will be applicable, and different probabilities assignable. It is the same with the modal words 'possible' and 'necessary'. We can say on Tuesday, 'It is possible that it will rain tomorrow', but on Thursday, we can no longer say 'It is possible that it rained yesterday'. Or, to take Cooper's example,‡ we can say 'It is possible that Higgins is dead' if we were prepared to describe Higgins as a man in a house on fire, but not if we were prepared to describe him as a man rescued by a fire-escape. The difficulty arises because in these everyday contexts we can refer to particulars directly without describing them, and therefore without having to decide under which description we are referring to them, and hence within which universe of discourse the propositional function is to range. 'Smith', 'Higgins', 'this', 'tomorrow', 'it', refer without specifying the universe of discourse, the subject of conversation. "Will *what* rain?", "Will *what* come up

† W. V. Quine, *From a Logical Point of View*, Cambridge, Mass., 1953, Ch. VIII, pp. 142–59.

‡ Neil Cooper, "The Concept of Probability", *British Journal for the Philosophy of Science*, XVI, 1965, pp. 228–9.

double six?" With material objects and with public events it often does not matter how we specify them, for many different specifications will identify the same material object or the same public event equally well. It is enough simply to refer to the material object or public event, so long as our referring phrase enables our hearers to identify the material object or public event we have in mind. But propositions—which are what probabilities attach to—are importantly un-thing-like in that a different specification characteristically specifies a different proposition. Although the event referred to in "It will rain tomorrow" is the same whether uttered at 6 p.m. or just before midnight, the correct specification of the proposition to which we are assigning a probability may be different; in one case $F(a)$, in the other $F(b)$, where a and b represent the different meteorological descriptions of Wednesday available to us at 6 p.m. and 11.58 p.m. on Tuesday, under which it would be reasonable to think of Wednesday when wondering whether it would rain.

Material objects and public events are *substances*. They exist independently of us, and are unaffected by how we know them, or whether we know them at all. In talking about them we do not have to specify them, but only to refer to them. Any reference that suffices to identify them is enough, and often a small, and minimally informative label is adequate for the purpose. We tend to think of ourselves as labelling substances in order to talk about them, but of the substances themselves being "there" all the time whether or not we stick labels on them, and not altered and certainly not constituted, by the labels we attach. But the universes of discourse over which propositional functions range are not substances, and are affected by how we describe them and refer to them. And so singular propositions puzzle us, because they seem to be straightforward propositions about straightforward public events, but, when candidates for probability-values, are regarded as propositional functions and therefore not about substances at all.†

† See further below, Ch. XII.

VII

EQUIPROBABILITY AND RANDOMNESS

THE Classical Theory of Probability was very different from that developed here. It began as a Theory of Games—games of chance with dice or cards. There was always some basic assumption of Equiprobability: that the chance of a die falling with the face marked six uppermost is the same as the chance of its falling with a face marked with five uppermost, *etc.*; or that the chance of drawing any one particular card from a pack is the same as that of drawing any other particular card. Once such an assumption is granted, we can calculate the probabilities of more complex propositional functions, such as scoring more than twenty-one with six throws of a die, or of a hand being dealt containing thirteen cards all of the same suit. Sometimes considerable logical acumen is required to attack the problem in a way that will yield a neat solution. Most of the work in the Seventeenth and Eighteenth Centuries was of this sort; and it is still often required for modern problems concerning the size of telephone exchanges or number of exits required in a supermarket. The assumptions required by the Equiprobability approach are open to question, as we shall see: but it remains the best approach for beginning to grasp the concept of probability, and in particular for being able to manipulate it. It lies outside the scope of this book to give examples or exercises: but the reader who wants to learn how to use the concept can do so only by practice; and the classical approach is the best to begin with.†

The philosophical difficulty lies in the apparently *a priori* assumption of equal probabilities. We assume that a die has a $\frac{1}{6}$ chance of falling on any one of its six faces; that a coin has $\frac{1}{2}$ a chance of falling heads and $\frac{1}{2}$ a chance of falling tails; that there is a $\frac{1}{52}$ chance of drawing any particular card out of a pack. The argument in each case is the same: it is based on an assumption of

---

† See, W. A. Whitworth, *Choice and Chance*, 5th ed., Cambridge, 1901 and W. Feller, *An Introduction to Probability Theory*.

equiprobability. We argue that each face of the die, each side of the coin, each card of the pack, has the same chance as any other. If we are pressed further on why each of these should be equiprobable, we are at a loss what to answer: it seems so obvious that they are, we cannot see how it can be doubted: and so we fall back on a self-evident intuition. We claim that it is *a priori* obvious that they must be equiprobable; and thereupon run foul of empiricist-minded philosophers who cannot abide it that *a priori* premisses should yield synthetic conclusions.

We can discover what is at stake by the common philosophical procedure of asking what one would say if things went wrong: in this case if the observed frequencies ran counter to what one would have expected from the *a priori* probabilities. If, for example, a die showed up six one quarter of the time. The answer is obvious. We should say that it was not a true die, but biassed in favour of six. Indeed with tossing a coin, heads and tails do not come down equally often, and the real probability of a penny coming down heads is not 0·5000 but something more like 0·5287.†
Dice do in fact come down with equal probabilities for each face, but only because the manufacturers test them, and do not allow biassed dice to reach the market. Thus in part our *a priori* probability is a mixture of an analytic truth and a synthetic *a posteriori* one. It is analytically true that an unbiassed coin, a true die, a well-shuffled pack, will produce each possibility with equal probabilities: it is synthetically true that the actual devices on the market do work and do the job they are required to do—if they did not, we would not use them, but would look for some other device.

This short answer does not completely exhaust the problem. For it *does* seem obvious that a die *ought* to have an equal probability for each face. It is so symmetrical that it could not, we feel, have any reason for falling on one rather than any other. We can see this more clearly if we suppose that we found a die to be biassed. We should then *expect* to discover something in its physical structure—say, a concealed off-centre weight—which would explain its asymmetrical behaviour. And if we could not find anything, we should be very worried indeed, and go on and on looking for something, and feeling that there *must* be some-

† I am indebted to my pupil, Mr R. C. Bennett, of Merton, who, with other undergraduates, tossed a penny a large number of times to obtain this result.

thing to explain the anomaly. So too with the coin and the pack of cards. The principle of insufficient reason operates in the sense that we should demand an explanation of any other than equal probabilities, but are quite happy to accept equal probabilities without demur. It is an *a priori* presumption rather than an *a priori* certainty. We could be wrong, but until the evidence shows us to be wrong, we shall assume *a priori* probabilities, and if the evidence does show us to be wrong, we shall look for an explanation. But why, it now may be asked, should we assume anything *a priori*? The answer lies in certain general ideas we have about nature. Thus with the die and the coin we invoke certain principles of symmetry and certain theories of what can and cannot be a cause in mechanics. We feel that the die and the coin have axes of symmetry as far as their mechanical properties go, and that the differentiating characteristics of their faces are causally irrelevant. In each of these beliefs we could be wrong. Indeed, in the case of the penny, we are wrong: the characteristics on the obverse and reverse sides of a coin are not causally irrelevant, and produce a slight bias in favour of heads. In other cases, if we cannot find an asymmetry, we shall posit it. If, for example, a needle is dropped from a height on to the floor, we should expect it to be as likely to lie in any one direction as any other. If, however, this proved not to be the case, if, for example, it lay along a North–South line very much more than an East–West one, we should look for an explanation—in this case we might suspect that the needle was magnetized, and was being affected by the earth's magnetic field, which is not symmetrical. So too, when neutrons did not come out of nuclei in every direction equally, it was evidence for a hitherto unsuspected asymmetry in the structure of the nucleus; it was, in fact, the experimental proof of the non-conservation of parity. But this was not an expected result. The expected result was that there was no asymmetry involved, and that therefore the neutrons would have an equal probability of coming out in any direction. The principle of equiprobability is not a principle but a presumption; not a principle of indifference, but a demand that differences, if there are any, shall be accounted for. Magnetic and electrostatic fields were discovered because of the non-random patterns mechanical systems showed without there being any mechanical explanation for them. From the obvious symmetries of the die or the roulette wheel, we assume that there are no asymmetries involved; but with the proviso that if the results are not random, some explanatory asymmetry will be looked for. We save symmetry in the face of unequal probabilities by positing unknown forces. And so conversely, we can argue by *modus tollendo tollens* if we posit, by the principle of insufficient reason, an absence of unknown forces: if we are to accept symmetry we must accept equiprobability also.

The principle of insufficient reason sounds bad. Even if we call it only a presumption, it still is objectionable. How can one base any conclusions on ignorance? If we do not know anything, we do not know it, and cannot pretend that we do. Besides, we have only to formulate the principle whereby we extract knowledge out of ignorance, and we shall be able to derive contradictions from it. How then can we possibly even presume on the basis of ignorance? The answer lies on the dialectical nature of argument. The insufficient reason does not fail to suffice in a vacuum, but against an already established, although not conclusive, argument which it fails to overturn. The ignorance is not absolute. It is not the case that we know nothing of the properties of a die when we know of no reason why a six should come up any more often than a one. We know a lot about the mechanical properties of material objects in space, and we believe a good deal more: we have a theory of the isotropy of space, well established although not beyond dispute, which gives our assumption of symmetry *prima facie* reasonableness. It can be upset, as we have seen: but if there is insufficient reason to upset it, there is good reason to accept it and apply it in the particular case.

Equiprobability, and more generally, the whole concept of *randomness* is correlative with that of explanation. Events are random when they cannot or could not be explained.† A selection has been taken completely at random when there is no one who can or could say why just this sample was selected and not some other.‡ A sequence is random when there is no rule saying what the next member is to be.§ Bombs are said to have been dropped

---

† Compare W. Feller, *An Introduction to Probability Theory and its Applications*, Vol. I, New York, 1957, p. 40. "In *testing randomness*, the problem is to decide whether a given observation is attributable to chance or whether a search for assignable causes is indicated."

‡ See further below, Ch. IX, pp. 170–2.

§ See above, Ch. V, pp. 98–9, and below, pp. 118–19.

at random when they appear not to have been aimed at any target. The notion of randomness is thus a negative one; and that half accounts for the difficulty that has been found in defining it. Many writers have started by giving a negative definition, objective disorder,† the measure of our ignorance,‡ or something in terms of irrelevance,§ but have felt themselves impelled to go on and say something more positive; and that has been their undoing. Negative definitions are negative: it is a mistake to look for the presence of some feature where only the absence of another is specified.

The definition of randomness is made doubly difficult, however, by the fact that the concept of explanation, which randomness is the negation of, is itself a systematically ambiguous concept. There are as many different types of explanation as there are different sorts of answer to the question 'Why?' Aristotle was right in distinguishing different "Causes", or as we might less misleadingly term it, different "becauses", but he did not distinguish enough. There are many different answers 'because . . .' to questions 'why . . .?', and an inability to give an answer of any one type can be expressed by an ascription of the corresponding type of randomness.|| Thus we can talk of the configuration of the planets being random, notwithstanding the complete explicability of their movements in terms of Kepler's laws or Newtonian mechanics, meaning by 'random' that there is no "final cause" of their being where they are, no "rational" or "teleological" explanation in terms of volitions of conscious agents.¶ Or again, we talk of demand for, say, taxis or telephone lines, being random, not intending thereby to deny rationality to each passenger or subscriber, but only to deny that there was any pattern in the overall demand. Many philosophical puzzles have been generated by assuming the concept of randomness to have a constant and fixed meaning which cannot change from context to context. Thus Ayer argues for determinism, by saying that "if it is a matter of pure chance that a man should act in one way rather than another, he

† K. R. Popper, *Logic of Scientific Method*, London, 1959, p. 478.

‡ H. Poincaré, *Science and Method*, tr. Francis Maitland, London, 1914, Ch. IV, p. 65.

§ J. M. Keynes, *A Treatise on Probability*, London, 1921, pp. 287–91.

|| See further, J. R. Lucas, *The Freedom of the Will*, Oxford, 1970, §§7, 11.

¶ Compare J. M. Keynes, *op. cit.*, pp. 281–2.

may be free but he can hardly be responsible",† and later makes the same point using the word 'accident', saying that if it is not an accident that I choose to act as I do, then there is presumably a causal explanation of my choice, and that if it is an accident then I am not responsible. The words 'chance' and 'accident' here have the same logical grammar as 'random'; and are being used to indicate the absence of a causal explanation and the absence of a rational explanation indifferently. If a man cannot on reflection give reasons for his actions, then we may say that he was irresponsible, and that actions for which no putative rational justification could be offered are barely actions at all, but arbitrary reactions not susceptible of moral evaluation. If we cannot give a complete causal explanation of a man's activities, then we may conclude that his behaviour is not determined in the determinists' sense of the word. But unless one is prepared to argue that the word 'explanation' is being used in exactly the same way in both cases, it is clear that the absence of one sort of explanation does not necessitate the absence of the other; and that it is logically possible for an event to occur by accident, by chance, or at random, in the sense that there is no complete causal explanation of it, although it was intended and can be explained as a rational action, perhaps as part of a deliberately designed policy; just as it is equally possible for an event to be an accident in the ordinary sense, that nobody intended it to happen, though susceptible of a complete causal explanation none the less.

These three examples show both the systematic ambiguity of the concept and the importance of specifying the question to which no answer could be given and on account of which therefore an ascription of randomness was made. Whenever the words 'random', 'chance', 'accident', or 'coincidence' crop up, the philosopher should be on his guard. Until the underlying question has been framed, possibilities of confusion remain, and, with our normal elliptical mode of speaking, abound. It is very easy, unless one has tied oneself down to an exactly specified question in advance, to make unconscious shifts in the course of an argument or experiment, and extract quite remarkable conclusions

---

† A. J. Ayer, "Freedom and Necessity", *Polemic*, 5, 1946, p. 38; reprinted in *Philosophical Essays*, p. 275. For a later version of Ayer's argument see A. J. Ayer, *The Concept of a Person*, London, 1963, p. 254.

from quite unremarkable data. I believe this to be the case with many of the disputes about psychical research.

Since the concept of randomness depends on that of explanation, a further important consequence follows. Explanations are covertly universal. They are in terms of features or characteristics of events or situations, not in terms of the particular instances alone. It is true not only of causal explanations, as Hume dimly apprehended, and moral justifications, as Kant proclaimed, but of historical, psychological, legal and all other explanations as well. In all cases, if we are to give any explanation at all, we have to begin by pointing out general features. We may not be able to characterize a particular completely, and in the humanities at least, it is not expected that we should; and we are entitled to leave an unarticulated remainder in our description, and even fall back on it in the course of argument, and draw out of it further features now seen to be relevant,† but it is always further features and characteristics that we adduce: explanation is a matter of types, not instances, of universals, not particulars.

It follows that non-explanation is a matter of universals, characteristics, or features, too. It is not enough to refer to an event, and say that it is a random event: we must refer to it by means of a description, and as the description varies so will the randomness, even though the reference remains unaltered. The best examples are those where a sequence of numerals is used for labelling purposes—car registration numbers, railway engine numbers, telephone numbers. If I see a car registration number FC 1, I tend to attach special significance to this, and suppose it to belong to a rather important person. Equally I should comment on a road number B1066, or a telephone number 1234 or ABBey 2222 or Oxford 87654. We might argue that there was only a one in ten thousand chance of a four figure number being 1066, or of its being 1234, and therefore it was a most remarkable coincidence that this should have been the case: but then we reflect

† See further, J. R. Lucas, "The Lesbian Rule", *Philosophy*, XXX, 1955, pp. 195–213; and "Causation", *Analytical Philosophy*, first series, ed. R. J. Butler, Oxford, 1962, pp. 35–7. Probability statements are universalisable too, because they claim to be based on reasons; see above, Ch. I, p. 8, and also D. H. Mellor, "Chance", *Proceedings of the Aristotelian Society*, Supplementary Volume, LXIII, 1969, §7, p. 31. The difficulties about explanation, randomness and coincidence in this and the next chapter are like those about singular propositions in the last chapter.

that it would be equally long odds against any other four figure number turning up, say 7548, and yet we do not feel inclined to make anything of that. Or, to take another example, if when the cards are dealt, four bridge players each find that they have a complete suit, they would think it a near miracle so great are the odds against it. But the odds against any other "hands"of cards are equally great, and yet these are accepted every day without comment.

The answer in each case is that the number, or distribution of cards, is not just the one particular number that it is, but also can be described in a way which is both impossible for most other numbers and relevant to a putative explanation. On the London telephone exchange there are millions of conceivable telephone numbers but only seven at most in which all the digits are the same, and to have all the digits the same is highly memorable and might well be the result of deliberate choice. 1066 belongs to the small class of highly memorable dates in the way that 1027 does not. It is an adventitious significance which we wish on the road from Bury St. Edmunds to Long Melford in Suffolk that is the proud possessor of this number, and we do not believe there is anything in it, but we are all of us potential numerologists, and cannot help noticing where numerological explanations might be looked for even though we decline to look for them. Under the rules of bridge there is only one distribution of cards out of all the possible ones in which neither the composition of the separate hands nor skill in bidding nor skill in playing will avail any player anything; or more generally and more intuitively, there is a natural simplicity in the division of a pack by suits which is simpler than any other distribution of cards. It is the simplest distribution, and therefore suggests the possibility of there being some sort of explanation, just as FC 1 is not just one among many car regis-tration numbers but is the *first* one of all the FC ones, and there-fore might well have been assigned deliberately. When we ask whether this distribution of cards or this registration number was assigned at random, we have in our minds the unformulated question, "Why was this simplest distribution of cards assigned the players?" or "Why was this first of the FC registration num-bers assigned to this car?", and the distribution of cards and the registration number are being referred to under these descrip-tions. If we refer to them under other descriptions, the odds are altered. Similarly in other cases, if we alter the description, we alter the odds. What descriptions we can apply depend on what possible explanations we have in mind. The odds against any particular vehicle having the registration number NFC 57 and no other are enormous. But we do not think that the description "having the registration number NFC 57 and no other" could fit in to any possible explanation, and so we say that this is just a registration number at random and not one to which one could attach any significance.†

Bertrand's paradox arises from neglecting the negative definition of randomness.‡ Bertrand asked what was the probability of a chord chosen "at random" in a circle being longer than the side of an equilateral triangle inscribed in the same circle, and showed three different answers could plausibly be argued for, according to how the chord was chosen "at random". If it was by choosing two points on the circumference to be the end-points of the chord, we should give an answer of $\frac{1}{3}$: if it was by choosing one point on a radius to be the mid-point of the chord, we should give an answer of $\frac{1}{2}$: if it was by choosing one point anywhere within the circumference of the circle to be the mid-point of the chord, we should give an answer of $\frac{1}{4}$. These different answers arise, as Bertrand himself observed, because "a chord chosen at random" is itself indeterminate. We do not know what is *not* being done, because we do not know what is and what might be done. If we were using a roulette wheel twice over, we both know the general approach and the sort of particularity that is lacking: we can contrast this sort of randomness with using the roulette wheel once and taking the points at opposite ends of the diameter defined by the pointer. Similarly if we first choose a radius and then the point of perpendicular intersection between it and the chord, 'random' has a determinate sense, because with the general approach specified, it has a determinate contrast. Similarly again if we just choose a point in the circle which is to be the mid-point of the chord, we know what we are doing, and therefore what sort of results would surprise us and demand an explanation if we got them. If I was spinning a roulette wheel I should be surprised if it emerged that there was a greater probability of its coming to rest

† See next chapter, Ch. VIII, pp. 128–34, for further discussion.

‡ J. Bertrand, *Calcul des probabilités*, Paris, 1889, pp. 4–5; or see W. C. Kneale, *Probability and Induction*, Oxford, 1949, pp. 184–5.

in one direction than in another. If I was picking a point "at random" on a given radius, I should be surprised if points in any one part of the line had a greater probability of being picked than those in any other. And similarly, if I was picking a point "at random" in a given circle, I should be surprised if points in any one area within the circle had a greater probability of being picked than in any other. In each of these three cases the description is definite enough for us to tell what is being excluded: or better, we know what principle non-randomness would jeopardise—the isotropy of space in the first case, its linear homogeneity in the second, and its areal homogeneity in the third. A thing can be random only under a description which indicates what the non-explanation on offer is. Once we suppose randomness to be a positive quality which can be ascribed independently of descriptions and without regard to the sort of explanation that might be, but is not, forthcoming, we are involved in endless confusions and paradoxes.†

In probability theory we sometimes use randomness as a safeguard against von Mises' place-selection, sometimes in order to introduce a probability so as to facilitate computation and argument. An experimenter may use a random sequence in order to guard against his own choices introducing some systematic error into his experiments. It is particularly important when dealing with animals, who may learn any pattern of choice if there is one, or tell from the experimenter's attitude what his decision was. Schoolboys, asked the gender of words in Latin or French exams, will seek to discern some pattern in the questions, and after three words they know to be masculine will assume that the next one must be feminine. A general principle, therefore, is that to ensure arbitrariness, the experimenter does not choose between the alternatives under their own description, but tosses a coin or chooses cards from a pack or opens the Bible at random, and the result determines by some antecedently specified rule the action to be taken. In Fisher's celebrated example of the old lady tasting tea to see whether the milk was put in the cup before the tea was poured or after,‡ it would spoil the test if she was given the dif-

† See, for example, G. Spencer Brown, *Probability and Scientific Inference*, London, 1957.
‡ R. A. Fisher, *The Design of Experiments*, 7th ed., Edinburgh, 1960, Ch. 2, §§5–10, pp. 11–21.

ferent cupfuls alternately or all the pre-lactationist ones first and all the post-lactationist ones afterwards. To avoid experimental solecisms of this sort, and to avoid the tell-tale glint in the experimenter's eye, the order in which the cups are served may be determined by drawing four cards out of a pack of eight, marked one to eight. And whereas a causal explanation of the lady's guessing the experimenter's plans from the way in which the experiment is conducted cannot be ruled out of court, a causal explanation of the lady's guessing which cards the experimenter's hand was going to light upon can be confidently dismissed. In such cases the word 'random' is being opposed to 'design', 'conscious pattern', or 'causal law'. By introducing a random element we are ruling out the possibility that the results could be explained by the old lady's reading the experimenter's mind, or there being some other causal law which produced misleading results which the experimenter had not guarded against. We can defeat old ladies, intelligent chimpanzees and even some sorts of Laplacean demons by making our experimental moves unpredictable-by-them because selected by us through some randomising procedure. Any putative causal law—*e.g.*, that the old lady really can tell the difference between the two sorts of tea— which might otherwise have been apparently verified because she guessed the order in which the experimenter would serve them, will now be decisively falsified, in a traditional, non-probabilistic, way. A randomising procedure or a random sequence used for such purposes does not need to have any statistical properties. All that we require is that it should be effectively unpredictable, and too complicated to tally by coincidence with any plausible extraneous causal law.

Often, however, we introduce random sequences just in order to introduce probabilities for the sake of convenience and economy, particularly so when we are already dealing with continuous magnitudes.† If time and money were no object we could use Mill's Method of Difference. If we want to discover the quantitative influence of one factor, say A (*e.g.* the application of potash fertiliser) upon some quantitative effect, say Z (*e.g.* the yield of corn), and we suspect that B_1, B_2, . . . , B_r, may also be relevant we could conduct r, or better 2^r, pairs of experiments, to see exactly what effect the presence or absence of A will have in

† See below, Ch. X.

conjunction with any particular B, or better any particular combinations of B's. But it would be a laborious task, and often if the B's are thought to be fairly independent of A and of one another, the experimenter will do a number of experiments with A present and with A absent, and use a random sequence to ensure that the probability of each B's being present was the same for those experiments in which A was present and those in which A was absent. This secured, it is reasonable to add up the total yields for the experiments in which A was present and the total yields for the experiments in which A was absent, and compare them. The difference will be arguably due to the presence of A, since the contribution of each of the B's to both the totals is probably more or less the same, and therefore will not affect the difference between them. We have been able to avoid the labour of working through all the B's separately by using probabilities to quantify them, so that they could be manipulated like numbers, and subtracted as well as added: Boolean algebra has no inverse operation of subtraction,† and we can work Mill's Method of Difference only by examining each pair of cases separately—incidentally acquiring much information we do not need about the effect of the various B's on yields. Probabilities enable us to move from the Boolean algebra of presence or absence of factors to the arithmetical algebra of expectation of effect, and thus to concentrate all our attention on the difference of yield that can reasonably be attributed to the one factor we are interested in. It is a great economy.

The random sequences to be used in these cases need to have statistical properties above all else. It does not matter much if a person could guess what the next figure was going to be: but it would matter if in a number of tests the random tables led us to have significantly more experiments with a particular B present among those where A was present too than among the control group where A was absent. Whatever our procedure was, our experimental results would be vitiated if this condition was not satisfied. But to satisfy it raises difficulties. If we use a chance-device—say tossing a coin—there is a small, but non-infinitesimal probability of getting a run of, say, one hundred tails: and if this 1 in 10^{30} chance were to happen to us, we should be right to reject the results given by that chance device and use some other

† See above, Ch. III, pp. 25–6.

method giving a better balance. A similar problem besets the compilers of Random Sampling Numbers. Kendall and Babington Smith indicate the portions of their table (5 thousands out of 100) which it would be better to avoid in sampling requiring fewer than 1,000 digits.† But this is nothing terrible, once we realise that randomness is a negative concept. What features we require it to have depends on what we want to avoid. If it is statistical bias over relatively short runs we can avoid that, if over longer runs then that. Only, we cannot avoid both, because they are incompatible requirements, since it is part of our requirement for the long run that it should include a few short runs which deviate noticeably from the others. But once we have decided what experiments we are going to do, and for what purposes we are going to use random sequences, it is perfectly possible to select one which satisfies our negative requirements.

The assumption of equiprobability in games of chance was justified in that if the probabilities turned out to be unequal we should want an explanation, and any such explanation would not fit in at all easily with what we already know and believe about ourselves and the world around us. In some cases, as we have seen, there is an assumption of symmetry. But always there are other assumptions, and these can cause philosophical perplexity. If we accept the normal determinist mechanics which applies to the motion of solid cubes, we must believe that the question of which face of a die will lie uppermost is determined by the initial position and momentum and orientation and angular momentum of the die, together with certain conditions of friction and elasticity between die and table. Similarly, and more obviously, with a roulette wheel; the final position of the pointer must be determined by the initial push. The randomising devices, it would seem, do not really introduce a random element at all. A Laplacean demon could have predicted the outcome before the throw was ever made.

All this is true. But it does not follow that randomising devices do not work. It is partly a matter of specification. A Laplaceanly fully specified throw of a die would indeed be the subject not of a probability-judgement but of a true singular proposition: but we

† M. G. Kendall and B. Babington Smith, *Tables of Random Sampling Numbers*, Cambridge, 1939. See M. G. Kendall and A. Stuart, *Advanced Theory of Statistics*, London, 1958, Vol. I, p. 218.

do not—cannot—specify so fully. To put it another way, it would be sufficient for many purposes that the randomness produced by our randomising devices should be an unpredictability-in-practice rather than an unpredictability-in-principle. It may be that a Laplacean calculator could work out from the way I held the dice-box which face would be uppermost. But, mediaeval fables apart, I do not play dice with fiendish calculators full of malevolence towards me, but with other limited mortals, who can only describe the situation in the same terms as I do. Here it is a question of imperfect information.† We are, at least metaphorically, betting on the outcome of particular throws, and if we could specify the situation more completely, we would: but then of course equally, if other men could do this, we should no longer use dice in games of chance. It is a synthetic truth that we cannot in fact predict how coins, dice and roulette-wheels will behave in particular cases: but it is an analytic truth that in games of chance the procedure must be describable by each player only in probabilistic terms.

Nevertheless some metaphysical disquiet remains. Unpredictability-in-practice may be enough for games of chance, but often we posit an absolute unpredictability incompatible with a Laplacean view of the world. We do believe, most of us, that the results produced by a randomising device are unpredictable-in-principle: not simply because of the construction of the device, but because it is operated by a human being. We are not prepared to replace the human operator himself by a machine, for then we should expect the die to be given the same initial conditions each time it was thrown, and so it would turn up the same way each time. Apart from machines embodying some electronic or radioactive principle, where there is much evidence and strong theoretical considerations supporting the expectation that we shall get some phenomenon occurring only with a certain probability and not otherwise determined,‡ we have no randomising *machines*: all randomising devices have to be operated by a human operator. They do not introduce the element of randomness in this sense, but rather canalise it.

Determinism is usually regarded as a question for moral philosophers, and many of the arguments against it lie outside the

† See further below, Ch. XI.
‡ See below, Ch. XII.

scope of this book. But at one point we invoke human freedom in our philosophy of science. It is only on the assumption that human beings are not themselves machines and are capable of arbitrary choices and can introduce a random element into natural phenomena, that the method of experiment will establish the existence of natural laws. For, in order to establish the existence of a natural impossibility or natural necessity, we need to be able to vary the conditions arbitrarily so as to put the putative impossibility or necessity to the test. In practice it is enough that our interventions should be independent of the phenomena under investigation, and we have practicable criteria of independence. But on the Laplacean view there are no independent interventions, since everything is determined by the antecedent state of the whole universe; and if we do not believe that our experiments are arbitrary and independent interferences with the course of events, but are determined, antecedently to any decision of ours, by the state of the whole universe, then our experiments are not real experiments, and do not rule out, as they otherwise would, the possibility of the whole sequence of observations being "rigged" so that it gives the appearance of causal connexion without there being any causal connexion really. We can only discover the necessities of nature against the foil of the non-necessity of our own actions. Therefore, we must regard human beings as undetermined, that is as introducing randomness, so far as causal regularities are concerned, into the course of natural events.

If we regard human beings as introducing randomness into nature, we may ask what need there is for randomising devices. The answer is obvious. A human being can act in a way that is causally random—not completely explicable in causal or determinist terms—but not inexplicable. In particular, if a player wanted to win, his putting down the die with the six uppermost would be entirely explicable, though no complete causal account could be given of why he placed the die in the exact place that he did. But if he throws the die, having only a limited and non-Laplacean intelligence, he is unable to predict where and how he should throw the die in order to bring the six uppermost. The randomising device blocks the possibility of a teleological explanation. All the player can do is to choose some starting position and method of throw without knowing what the result of that choice will be. There are a very large number of significantly different starting points and methods —we have reason to believe that a very small variation in initial conditions would often alter which face came up—and, provided the die is symmetrical, and subject to certain conditions of continuity and differentiability, these are equally distributed among the six possible outcomes.† The player in effect draws one of a large range of starting points and methods, which, through the mechanism of the die, has one of six labels on it, which he cannot know until after he has made his choice. Drawing cards from a pack, face downwards, is the purest case of a randomising device. The function of the others, the die, the coin, the roulette wheel, is, so to speak, to keep the face downwards, to guarantee that the results are teleologically as well as causally random.

We are making here a second assumption about ourselves, that our power to influence events is limited to certain well-known methods, and that when we cannot use these—as when we draw a card from a pack or shake a die—we cannot impose our wills upon the outcome at all. If when we drew a card from a pack we were more likely to draw an Ace than a Ten, or a Red than a Black, we should want to know why; and we are already firmly convinced that any such regularity, unless it could be explained by cheating or with reference to the texture of the cards, or in some other humdrum way, would accord ill with what we already know from innumerable observations and much experience about our ability to influence cards and other things. We do not believe in psycho-kinesis. It is partly an *a posteriori* empirical truth—we often have willed things to happen and they seldom have, unless we have taken steps actually to do something to bring about the desired result. It is also partly an *a priori* conceptual truth—if merely to want something was enough to make it happen, we should have to be very careful what we wanted. At present we can want freely, because we are answerable only for our actions: but if wishes were effective, they would be weighed down by the burden of responsibility, and we could never let them fly freely to the sky. Again, at present, there is a

† See H. Poincaré, *Science and Hypothesis*, London, 1905, and New York, 1952, Ch. XI; or W. C. Kneale, *Probability and Induction* Oxford, 1949, §30, pp. 142–4. The argument will not bear as much weight as Kneale wants to rest upon it. See D. H. Mellor, *Chance*, Cambridge, forthcoming.

certain privacy of intention: only when I translate my intentions into action do they become public property, and only over overt behaviour can disputes with other men arise: but total psycho-kinesis would dissolve this barrier between my private decision-making and public criticism and conflict, and if what I thought in my chamber was automatically brought into effect, I should be having all my private thoughts proclaimed on the housetops. Our whole concept of individual personality requires that psycho-kinesis should not be a normal phenomenon; and though we cannot say that it never happens, we are entitled to say that it cannot often happen, and that there is *a priori* presumption against it, and therefore in favour of Equiprobability in the drawing of cards.

The principle of insufficient reason is thus tenuous and complicated, but valid. It depends on several assumptions: sometimes symmetry, sometimes continuity, in the world around us; and in ourselves, freedom and non-omnipotence. These assumptions do not all hold always, but it is reasonable, in the absence of evidence to the contrary, to assume that they do. And then there follow conclusions of Equiprobability.

VIII

INVERSE ARGUMENTS

THE argument from Equiprobability constitutes one way of assigning definite numerical values to certain propositional functions. It is not, as some modern critics have alleged, totally fallacious: but it is sometimes tenuous, and often inapplicable, especially in the biological and social sciences, and by itself does not provide an adequate foundation for a theory of statistics. We depend on other arguments to lead us from non-probabilistic premisses to probabilistic conclusions, notably Bernoulli's Theorem† and Bayes' Theorem. Of these, Bernoulli's Theorem is, I believe, the more fundamental, but Bayes' Theorem has attracted more controversy.

Inverse arguments cause difficulty in probability theory because the Law of Contraposition and the Rule of Inference *modus tollendo tollens* do not apply. In ordinary non-probabilistic discourse it follows from 'If he is an uncle he is not an only child' that 'If he is an only child, he is not an uncle'; and similarly if we have the premiss 'If Hannibal marched on Rome, he took it' and deny the putative conclusion and say instead that 'Hannibal did not take Rome', then we can infer that the antecedent of the hypothetical is false; that is, we can say 'Hannibal did not march on Rome'. In the propositional calculus we express the Law of Contraposition by the equivalence

$$q \supset r . \equiv . \sim r \supset \sim q$$

and the Rule *modus tollendo tollens*

$$q \supset r$$
$$\frac{\sim r}{\therefore \ \sim q}$$

These enable us to argue backwards as well as forwards, and it is

† See above Ch. V.

merely a matter of convenience which way we articulate our arguments. Probability theory is different. If we know that a man of twenty who smokes 40 cigarettes a day has a 30% chance of dying of cancer of the lung by the time he is forty, we cannot infer that an evidently alive man in his forties probably did not smoke 40 cigarettes a day when he was twenty. For our original statement was about twenty-year old cigarette-smokers, and does not tell us enough about live forty-year olds for us to make any statement about them. To take an extreme example: it is very improbable— say, only 1% probable—that a twenty-year old man who smokes 40 cigarettes a day will live to be ninety; but among ninety-year olds there may be a lot of Churchillian smokers because if a person was tough enough to live that long he would have to be so tough that not even smoking could hurt him. Our symbolism expresses the difference between what the probability-judgement is about—twenty-year old 40-cigarette-a-day-smokers—and what it is being said about them—dying-of-cancer-before-they-are-forty—by representing the former by small letters $a, b, c, \ldots,$ etc., and the latter by big ones, $A, B, C, \ldots,$ etc. Whitehead and Russell's propositional calculus does not, in effect, make this distinction between subject and predicate: 'All ravens are black' is rendered

$$(x).\, \mathrm{Raven}(x) \supset \mathrm{Black}(x)$$

which immediately contraposes to

$$(x).\, \sim \mathrm{Black}(x) \supset \sim \mathrm{Raven}(x).$$

I should claim it in general as a great merit of using Smiley's many-sorted logic rather than Whitehead and Russell's that it does not open the door to contrapositional nonsense about non-blacks being non-ravens or the immortals on Olympus being non-smokers; more especially in probability theory where *modus tollendo tollens* characteristically does not work.

Although *modus tollendo tollens* characteristically will not work, we want it to. Our strategy is therefore to move by means of Bernoulli's Theorem from the normal ranges of probabilities where it does not work at all to the special extremes where it almost does. I say 'almost' advisedly. Although if it is highly probable that an Oxford man knows Latin, it is reasonable to infer that a man who does not know Latin has not been to Oxford, the argument is not watertight, as the Churchill example shows. We may be able to explain why, although it is very probable that a g is F, it is not at all probable that something which is not F should not be G. But we do need to explain it away, if we are going to refuse the inference. As the probabilities involved become nearer and nearer truth and falsehood, so it would become more and more unreasonable to refuse, without reason, to argue by *modus tollendo tollens*: for, as we have seen,† it is not just the extreme limits themselves but their neighbourhoods which are to be identified with truth and falsehood: and although what constitutes the neighbourhood is a matter for discussion and depends on the context of argument, to insist on regarding not the neighbourhoods but only the limit-points themselves as representing the values True and False, is to opt out of probabilistic argument altogether.

We have opened up a possible line of argument in which we keep near to the extremes of truth and falsehood, and so can use *modus tollendo tollens* to argue in reverse by Bernoulli's theorem from non-probabilistic premisses to some sort of probabilistic conclusions. There are, however, three further difficulties to be surmounted: we have still to elucidate the concept of coincidence; there is an element of arbitrariness in our identification of truth with *a* neighbourhood of 1, and falsehood with *a* neighbourhood of 0; and not all assemblies of statistical evidence will yield probabilistic conclusions.

Coincidences are not just improbable events. Philosophers who think that they are, fall into the Dover fallacy:

> There was a young curate of Dover
> Who bowled twenty-five wides in an over,
> Which had never been done
> By a clergyman's son
> On a Tuesday in August in Dover.

In a similar vein it is easy to argue that *any* result of ten tosses of a coin, *any* sequence such as

$$\text{HHTHTHHTTH}$$

has, on the hypothesis that the probability of heads is $\frac{1}{2}$, itself a probability of $2^{-10} = 1/1{,}024$, which is less than even the $0{\cdot}1\%$ significance level, and should therefore lead us to reject the hypothesis without more ado. The mistake lies in not recognizing

† In Ch. V.

that coincidences are always *just* coincidences or *mere* coincidences. The word is being used to say not simply that something
is the case, but, more importantly, that something else—something more—is *not* the case. It was just a coincidence that I met
an old friend in St. James' Park—we did meet, but *not* because we
made a *rendez-vous, nor* because I had thought that there would
be a good chance of my meeting him or some other of my friends
in the Civil Service if I walked there at lunch time, *nor* because he
knew I was coming to London and would be likely to be walking
there. Although in saying that it was a coincidence, we are saying
that we did meet, and are allowing that there were separate explanations of our each being in St. James' Park on Wednesday,
July 17th 1968, we are primarily concerned to deny that there was
a special explanation of our both being at the same place at the
same time. Coincidence is a *dialectical* concept. Its place is in the
context of a dialogue or argument, to reject an explanation which
the other person might reasonably posit. Like randomness, it depends on description and obtains its colour from the surrounding
context of explanations otherwise to be expected, and the same
event may be a remarkable coincidence under one description
and quite unremarkable under another. As with other singular
propositions,† we are easily misled by the fact that we can suc-
cessfully refer to a particular event under many different descriptions, but would not assign the same probability to the different
propositions incorporating the different descriptions: only, in the
case of coincidence, the relevant description varies not with time
or taste, but putative explanation to be rejected.‡

Coincidence plays a central part in probabilistic arguments
because we can use Bernoulli's theorem to reject, but not directly
to establish, hypotheses about probabilities. We can argue for a
hypothesis only by arguing against others. And so it is only by
considering some sort of alternative hypothesis or explanation
that we can obtain any purchase on our argument. We are,
essentially, Popperites. We can refute, but cannot directly prove.
We need to have a hypothesis to refute, in order to make any
progress at all. And therefore we need an opponent, actual or
imaginary, to put up a counter-thesis for us to demolish. We ask,

† See above, Ch. VI.
‡ For another example and illuminating discussion, see Michael Polanyi,
Personal Knowledge, London, 1958, Ch. III, §1, p. 33.

sometimes seriously, more often rhetorically, 'Well, what alternative do you suggest?', and then in the same breath show that to ascribe the results to the alternative would require a coincidence too strange to be acceptable. Just as probability itself is at root a dialectical concept, so probabilistic arguments ought to be understood as dialogues. Dialogues do not fit into one standard pattern, and often cannot be fitted into some schema of formal inference valid in monologous discourse.† *Prima facie* arguments are used which are essentially open to rebuttal, and are to be accepted only in as much as they have not in fact been rebutted. Silence in the face of a rhetorical question must often do duty for proof. We argue by inviting rebuttal, and only when rebuttal is not forthcoming, and no reason is offered for supposing a coincidence to have occurred, do we then refuse to ascribe the phenomenon to mere coincidence.

When coincidence is invoked in an argument about probabilities, some *alternative* explanation is in the field, and we need to make explicit what the alternative explanation is, in order to know what the question at issue is, and so in what way the available evidence is relevant to giving an answer. The difficulty is that there are many hypotheses which we could refute, but which are not worth refuting. Many are non-starters, quite apart from any statistical evidence. We need to erect our cockshy carefully, so that it will not fall of its own accord, but will stand up enough to be knocked down by the evidence. Indeed, we must select the strongest challenger in the field and show that it is untenable in the face of the evidence if we are to make any progress in establishing our own explanation. How to pick the strongest is a matter of judgement as much as of following any formulated rules. It depends very much on the subject. Consilience is important. We have a vague idea on general grounds of what are possible causal connexions, what may be relevant factors, what form a causal law might take; and in the light of these we spot what the most likely alternative explanation is, and see whether or not it could at all easily have accounted for the phenomena that actually occurred.

The most likely alternative hypothesis is often, but not always, a "Null Hypothesis". I have ten bad oysters in a barrel of one

† See J. R. Lucas, "Not 'Therefore' but 'but'", *Philosophical Quarterly*, **16**, 1966, pp. 289–307.

thousand from one shop, and thirty in a barrel of one thousand
from another shop, and maintain that the second shop is less
good than the first; "how else do you account for it?" I ask.
"Perhaps the probability of an oyster's being bad is the same for
both shops," you answer "and it was *just luck* that one lot happened to be better than average, the other worse". This is the
null—no real difference—hypothesis. So we consider it: we suppose that the probability is the same, 2%, in both cases, and calculate the probability then of one sample having a proportion of
only 1% the other of 3%. It turns out to be less than one in
twenty, and therefore it would be quite a coincidence—although
not a chance in a hundred or a chance in a thousand—for there to
have been such a discrepancy.

Whatever the alternative hypothesis is, it will determine not
only what the question really is to which the statistical figures
yield the answers, but the canons of interpretation in accordance
with which we construe a particular set of figures as constituting
an answer. The way in which the alternative explanation determines the canon of relevance can best be illustrated by the Dover
argument that any sequence of heads and tails is as improbable
as any other (of similar length), and that any result whatever of
coin-tossing should count as much as any other against some probabilistic hypothesis. "By what right," the Dover philosopher
asks, "do we argue that the sequence

HHHHHHHHHHH

counts against the hypothesis that the coin is unbiassed, when we
do not regard the sequence

HHTHTHHTTH

as counting against it equally much?" The answer depends on the
Dover philosopher himself. "What alternative explanation do
you offer?" we ask; "there is an obvious explanation of our
getting all heads—that the coin is a two-headed one—and it
would require a considerable coincidence to displace that explanation in favour of a one-in-a-thousand chance. But how do you
explain the second result?" The question is not totally rhetorical.
It can be answered. Indeed, we can devise an answer which would
make relevant not only the proportions, but the exact order, of
heads and tails in the second sequence. We do not always regard

order as irrelevant. If, for example, a man from Duke University claiming psycho-kinetic powers had said in advance that he was going to will a coin to fall HHTHTHHTTH, then the relevant description of the outcome would be whether it did or did not conform to the prediction made; and if, as in this case, the order had been predicted, the order would be relevant: my counter-thesis, that the man from Duke had no psycho-kinetic powers, and the outcome was simply one of those that might result if the probability of the coin coming down heads was $\frac{1}{2}$ would already be looking extremely thin, requiring us to accept coincidence whose probability was less than $0 \cdot 1 \%$. In such a case order would be relevant; but in general it is not relevant because we have certain general ideas about the background against which our explanatory hypotheses are framed. We believe, with reason, that successive tosses of a coin are independent of one another, and that the probability at each toss is the same as at every other one. It is only when these conditions obtain that Bernoulli's Theorem applies. Under these conditions order is irrelevant. For conjunction is commutative, granted independence; and therefore when we are describing the outcome of a number of independent trials, we are right to ignore the order and consider only proportions. If, against the background assumption of similarity and independence, we want to determine whether the particular hypothesis that the probability of the propositional function has a certain value or has a value that lies within certain limits, then it would be irrelevant to include the order in the description of the outcome, and under such a description no outcome could tell for or against the hypothesis in issue, but all would be irrelevantly improbable.

Order is irrelevant when we are dealing with a number of independent instances of the same propositional function, to which a probability is to be assigned. But not all sets of propositions are independent, nor may we assume that every propositional function has a probability value. And if it is these conditions that are in question then order may be relevant. To take the simplest example, if we had a sequence HTHTHTHTHT we should strongly suspect that it was generated by some determinist alternator rather than that it was the result of ten independent tosses. HTHTHTHTHT differs from HHTHTHHTTH in that it suggests an alternative explanation making no demands on coincidence whereas, in most circumstances, HHTHTHHTTH does not. If we satisfy ourselves that the alternative explanation will not wash—if we toss the coin some more times and get no further alternations, or if we examine the coin very carefully and make sure there is no hidden mechanism, then we shall make the necessary drafts on coincidence, and say that the tosses were independent, and a 1 in 500 chance occurred of there being an alternation. And once we are satisfied that it must be a Bernoulli trial subject to the conditions of independence and similarity, order will be irrelevant.†

Nor is it only order that is irrelevant. Too detailed a description of the proportions will also lead to paradox. As we have already mentioned,‡ it would be highly improbable to have exactly five hundred heads out of a thousand tosses of a coin, but we certainly would not normally regard exactly five hundred heads and five hundred tails as counting against the hypothesis that the probability of heads is 0·5. The reason is that the available alternative hypotheses are normally also probabilistic, and will be no more—but, on the contrary, less—likely to yield the actual result. If the alternative explanation was that the figures had been cooked, then the fact that there were exactly five hundred heads and five hundred tails would indeed be significant. It depends on the alternative hypothesis what the relevant way of describing the results is. Essentially, we want to know whether the statistical evidence, taken together with any background information we have, is enough to warrant our rejecting the alternative hypothesis and re-instating our own explanation in sole possession of the field. A coincidence is what we could explain easily, but can better explain as a more or less improbable consequence of an alternative hypothesis. And in view of our own and the alternative hypothesis we divide the set of reasonably possible sets of statistics into two classes, those which can be explained by the alternative hypothesis—as well as by ours—without invoking too improbable a coincidence and those which only by an unacceptably impro-

† R. A. Fisher, "On the Mathematical Foundations of Theoretical Statistics", *Philosophical Transactions of the Royal Society*, A, CCXXII, 1922, pp. 309–68, gives a much fuller account of the conditions under which order is irrelevant. He introduces the term *sufficient statistic* for the evidence from which all irrelevancies have been extruded, but it is a term which many find confusing.

‡ Ch. II, pp. 18–19.

bable coincidence can be explained by the alternative hypothesis. We are choosing between hypotheses; and this determines the lie of the land.† But the hypotheses are not standing equal in our sight: we can use Bernoulli's theorem only to reject, and are therefore concerned with the question whether the alternative hypothesis has been proved false beyond reasonable doubt, and not which of the two fits the evidence best. The burden of proof is thus in favour of the alternative hypothesis, and it is only when the burden is discharged up to some acceptable standard, that the alternative hypothesis can be decently dismissed. What should count as an acceptable standard is, in part, arbitrary; and this determines not the lie of the land but the exact line between rejection and non-rejection.

The interplay between the arbitrary and non-arbitrary elements in settling what shall be an acceptable standard has caused

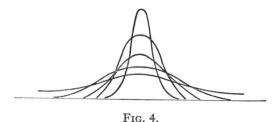

FIG. 4.

(Reproduced from p. 78).

considerable confusion. We cannot eliminate the arbitrary element altogether: but it would not be reasonable to pick on any arbitrary rule whatsoever as a rule for rejecting alternative hypotheses. What determines the lie of the land is Bernoulli's Theorem, which shows with what probability different frequencies are to be expected, on the supposition that the alternative hypothesis is true. Bernoulli's Theorem confirms the intuitive impression we had obtained that with the Binomial Distribution there is a greater and greater concentration around the middle as the number of terms increases, and that the proportion of the whole contributed by the middle (the "mountain" of figures 4 and 5) will

† Elsewhere in this chapter and particularly here, I am much indebted to a Fellowship Dissertation, *Towards an Objective Theory of Probability*, by Mr. D. A. Gillies, of King's College, Cambridge.

approach unity as nearly as we please provided we take a sufficiently large number of terms. Essentially, therefore, it divides the
possible frequencies into two classes, those in the mountain and
those in the plain; and the argument of Chapter V shows that
sooner or later we shall be justified in rejecting a hypothesis if
the actual frequency is one which on that hypothesis lies in the
plain. We may need to qualify this, in as much as we are hoping
to reject the alternative thesis in favour of a preferred one, and
there could be some results which, although in the plain of the
alternative hypothesis are equally much in the plain of the preferred one. The result

TTTTTTTTTTTTTHTTHTTTT

is very much in the plain of the hypothesis that the probability of
heads is 40% and of tails 60%: but I cannot use it to reject that
hypothesis in favour of the hypothesis that the probabilities of
heads and of tails are both 50%, since it would count against this
hypothesis even more; and it would be disingenuous to argue
from that result in favour of the 50% hypothesis as against the
heads 70%—tails 30% hypothesis, because although, if anything,
it favours the former, it counts against both so much that in de-
fault of other cogent considerations, we ought to reject them both.
Nevertheless, the division into mountain and plain is the essential
one, in spite of these qualifications. The typical case is one where
we have a result and an explanation which can easily account for
it, and we are wondering whether we can rule out some alternative explanation. And the question then is simply whether with
respect to that alternative hypothesis the result lies in the mountain or the plain.

The lie of the land does not tell us exactly where the plain
begins. The graph of the Binomial Distribution never becomes
flat, and however far we go away from the summit of the mountain, we can always find a more plain-like piece of plain by going
farther. But there are some guidelines. It would be absurd to
include in the plain any part closer to the peak than the *steepest*
part of the mountain. The points of steepest gradient therefore
give us an outside boundary for the plain. More naturally we
might reckon that the plain had ended, and at least the foothills
were beginning, when the gradient was itself increasing most
rapidly. These two qualitative features can be given numerical

precision in the limiting case where n, the number of cases, tends
to infinity, and the Binomial Distribution tends, as we shall
argue later,† to the continuous, Normal Distribution. The
Normal Distribution, as we shall argue, satisfies the differential
equation

$$\frac{\mathrm{d}}{\mathrm{d}x}f(x) = k \times f(x)$$

where k is a negative constant. It follows that

$$\frac{\mathrm{d}^2}{\mathrm{d}x^2}f(x) = \frac{\mathrm{d}}{\mathrm{d}x}(kxf(x)) = kf(x) + k^2x^2f(x)$$

$$= kf(x)[1 + kx^2]$$

and that

$$\frac{\mathrm{d}^3}{\mathrm{d}x^3} = \frac{\mathrm{d}}{\mathrm{d}x}[kf(x) + k^2x^2f(x)] = [k^2xf(x) + 2k^2xf(x) + k^3x^3f(x)]$$

$$= k^2xf(x)[3 + kx^2].$$

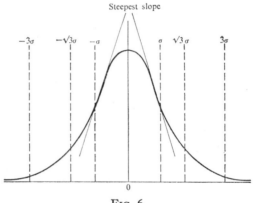

FIG. 6.

The points of steepest gradient occur when the second derivative
is zero. Since neither k nor $f(x)$ is zero, it must be when $1 + kx^2$
is zero; that is, when $x = \pm\sqrt{\frac{1}{-k}}$. Since k is a negative constant, $-k$ is a positive one, and so too $\frac{1}{-k}$, which we write σ^2.
The constant σ is a parameter of the curve, and is the standard
deviation.‡ The points of steepest gradient are thus where

† Ch. X, pp. 178–86.　　　‡ See below, Ch. IX, pp. 166–7.

$x = \pm\,\sigma$, that is, one standard deviation away from the summit in either direction. It is shown in figure 6; it is the point where the tangent crosses the curve, being above it towards the summit and below it away from the summit. In our unsophisticated thought we often take something like this to be the boundary between the mountain and the plain. In fact there is an approximately $\frac{2}{3}$ (68%) probability of being not more than one standard deviation away from the summit, and so only a $\frac{1}{6}$ probability of being more than one standard deviation away in a particular direction. In ordinary life we often regard this as being sufficiently improbable to be significant, although statisticians are more cautious about jumping to conclusions and regard the mountain as extending further. There would be quite a strong argument for saying that the boundary of the plain must be at least $\sqrt{3}$ of a standard deviation away from the summit, where the gradient is itself increasing

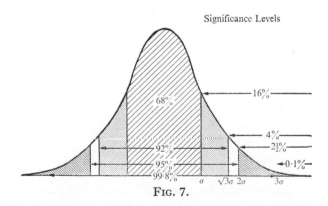

FIG. 7.

most rapidly, which would mean that there was less than 1 in 20, 5%, probability in being that far away from the summit in a particular direction; and we often go to 2σ or 3σ to be on the safe side. Figure 7 shows by different shading what the total probability of being within σ, $\sqrt{3}\sigma$, 2σ, and 3σ of the summit of a Normal Distribution, and (on the right-hand side) what the probability is of diverging from the summit *in that direction*.

Beyond this stage an essential arbitrariness appears. We have to *decide* what level of coincidence is to be regarded as insupportable. The argument of Chapter V showed that it was unreasonable to tolerate every improbability as being simply a coincidence but laid down no fixed level as *the* level: nor could it, for whatever level is laid down we can always construct an artificial case where it would be unreasonable to act on the assumption that coincidences beyond that level do not occur. All we can do is to consider the actual case, and decide what is reasonable in the particular circumstances of that case. It is a decision, a decision which leads, or could lead, to action, and therefore a decision which should be taken partly in the light of the action or inaction it may lead to. It would be entirely reasonable for me, a consumer, to change my oyster-shop on the strength of a 5% level. After all no great harm is done if I was wrong and it was a coincidence, and I can always take my custom back again. Moreover, life is short, and it is simply not worth spending much time on choosing the best oyster-shop. If I was a restaurateur whose livelihood depended on skilful purchasing, and particular if a change would involve upsetting long-standing contracts, I might be wise to wait and examine further samples before taking action. On the other hand, if I were not buying oysters but testing drugs, I should not dismiss the possibility of the "All-Clear" results being just a fortuitous coincidence until I had made sure that the probability of such a coincidence was much more slender—1 in 1,000, say, rather than only 1 in 20. For where life or health are at stake, we are rightly more cautious and rightly more prodigal of our time.

Three "significance levels"† are conventionally in use by statisticians: 5% (one in twenty), 1% (one in a hundred), and 0·1% (one in a thousand). They are not sacrosanct, but they are not silly. It would be perfectly all right to adopt 8·3%, 2·5%, and 0·5%; in unprofessional life we often regard 10% or 16% or even 20% as being too improbable a coincidence for us to bother about. Where we set the level depends on how much evidence we have, how much effort we are prepared to devote to obtaining fresh evidence, what sort of mistakes we most want to avoid and how much we want to avoid them, the general background of what sort of results could be expected, and the particular alternatives that we have in mind. So the conventional figures are not sacrosanct: but neither are they silly; for we must be prepared to fix *some* levels and stand by them, or we are not arguing seriously

† Also called 'size'; see J. Neyman and E. S. Pearson, "On the problem of the most efficient tests of statistical hypotheses", *Philosophical Transactions of the Royal Society*, A, CCXXXI, 1933, pp. 289–337.

at all. It would not be silly to use other significance levels than the conventional ones, but it would be silly not to use any at all, and therefore it is not silly to use the conventional ones in the absence of reasons for choosing others.

The fact that significance levels are arbitrary makes it easier for us to accept that no argument based on significance levels is conclusive. Since the concept of coincidence is a dialectical one, it is infinitely defeasible.† Since our argument is an argument from silence on the part of our opponent in the face of our more or less rhetorical questions, we are always open to a further 'but', for it is always possible that our opponent should answer our questions and adduce fresh considerations in favour of the alternative hypothesis. We throw a coin ten times and get ten heads, and conclude that the coin is biassed; "but" says our hearer "I have thrown that same coin hundreds of times, and have got as many tails as heads". We mate a black mouse with a brown one, and find all its seven offspring are black, and conclude that the mouse was homozygous:‡ "but" says our hearer "its mother was brown". In each case, further considerations have been adduced which rebut our probabilistic conclusion. Although by itself a 1 in 20, or a 1 in 100, or a 1 in 1,000, chance is to be rejected in the absence of countervailing considerations, it may have to be accepted if countervailing considerations can be adduced. It is unreasonable to ascribe a phenomenon—ten heads, seven black mice in a litter—to coincidence gratuitously rather than accept an obviously available explanation, but reasonable to do so if we have some further reason to suppose that it cannot be explained in the obvious way and must be a coincidence. Not only may there be some argument we had not thought of, but further experiments may yield results which put our previous rejection in question. The initial run of ten heads may be followed by a much longer one in which the proportion of heads and tails is approximately equal. Or, as we noted earlier,§ it is conceivable that we should come across a first page of the Plays of Shakespeare, followed by a page of gibberish, followed by the subsequent

† For an account of defeasibility, see H. L. A. Hart, "The Ascription of Responsibility and Rights", *Proceedings of the Aristotelian Society*, XLIX, 1948, pp. 171–94; reprinted in A. G. N. Flew, ed., *Logic and Language*, Series I, Oxford, 1951, pp. 145–66.

‡ See below, pp. 147–9.   § Ch. V, pp. 88–9.

pages of Twelfth Night intelligibly and correctly typed. The procedure of "going on" is open-ended, and never gives conclusive or definitive decisions. It is a merit of Braithwaite's theory of probability that his "rejection rules" are tentative and provisional, and a probability statement once rejected may have to be reinstated again:† although, for this very reason, Braithwaite's rejection rules cannot yield a satisfactory *definition* of probability.‡

Bernoulli sequences are special. Not every set of results can be construed as a Bernoulli sequence, or explained in terms of similar and independent probabilities. Von Mises' two conditions, of convergence and of randomness, are idealisations for infinite collectives of the rules of thumb we apply to finite sequences to tell whether to regard them as Bernoulli sequences. Either condition may fail. If the condition of convergence fails, and we find the limiting frequency is oscillating wildly, we then suspect either that we are not dealing with a probabilistic phenomenon at all, or that the successive cases do not have probabilities that are similar and independent. If the condition of randomness fails, we suspect that there is an underlying deterministic regularity which we had not previously noted, or that at least the successive cases are not completely independent of their predecessors in the way required for Bernoulli's theorem to hold. Thus HHTHTHHTTH, HTTHTTTHHH, TTHTHTTTHHT already shows a decreasing proportion of heads in successive segments of ten which might begin to make us suspicious. If the sequence were much longer, and the variations persisted over much longer segments—say 60% heads in the first thousand, 50% in the second, 40% in the third, we should be very hesitant to say that it was a simple Bernoulli trial, with a 50% probability of getting heads: we should suspect the coin, or whatever chance device it was, of wearing out, so that the probability was not the same at each successive throw. A similar counter affords half an answer to the favourite among paradox-mongers, which is a sequence modelled on the rule-generated sequence

TTHTHHTHHHTHHHHTHHHHHT . . .

† R. B. Braithwaite, *Scientific Explanation*, Cambridge, 1953, Ch. VI.
‡ I owe this point to Mr. D. A. Gillies, Fellow of King's College, Cambridge.

where after the first T there is no H, after the second T one H, after the third T two H's, *etc*. Because there is this generating rule, the sequence fails von Mises' condition of not being affected by place-selection. But it has a limiting frequency—0 for tails, 1 for heads—and so encourages us to see how the frequency in a von Mises Collective might also tend to limits of 0 and 1 in spite of having an infinite number of counter-instances. From this it is often argued that Truth cannot be identified with probability 1, nor falsehood with probability 0. But the argument holds only against the Frequency theorists. They must allow such a sequence to have a limiting frequency. But we should be very suspicious, and rightly so, if the frequency continually drifted towards an unattainable limit. If, whenever we undertook another putative Bernoulli trial, of say a thousand tosses, we obtained a smaller proportion of tails, we should again suspect that the coin was wearing out and the probability changing; the fact that von Mises' condition of convergence was—just—satisfied would not save it. Or we might conclude that it was not a probabilistic phenomenon at all. There is no reason why every phenomenon that is not deterministic must be probabilistic, or that every sequence for which we cannot find a generating rule must be a von Mises Collective. It is a demerit of some subjectivist theories which define probability negatively in terms of ignorance that they lead to this conclusion. But the number of sequences satisfying von Mises' two conditions must be small compared with those that do not, just as the number of sequences generated by a finitely specified rule is small compared with those that are not. There are many sequences, and hence there could be many sets of observations, which could not be explained probabilistically, not because they are too regular, and can be explained deterministically, but because they are too irregular, and not subject even to statistical regularity, and so not susceptible of any explanation at all.

Of course, in any of these conclusions we might be wrong. Just as with the alternating heads and tails,† so with apparently drifting frequencies, we cannot be sure that they are not the improbable but possible freak results which probability theory leads us to expect. All things are possible to chance. Whatever the initial segment of a sequence, we may always find that our imputations of improbabilism although justified on the evidence were

† See above, pp. 132–3.

falsified in the event: and that after straying far from the probable proprieties into highly improbable proportions, the frequency gradually retraces its steps to orthodoxy, and we come to reconsider our opinion of the initial segment, and see it not as a non-probabilistic phenomenon but as a remarkable coincidence after all. Coincidence is an elastic concept. It depends on the opposition how far it will stretch. We are right to reject a probabilistic explanation if it involves too high a degree of coincidence, but right, too, to withdraw that rejection if subsequent evidence shows that any other explanation would be even more far-fetched.

We have now detailed the argument whereby we are entitled to reject an alternative hypothesis, and have shown how, in spite of the difficulties in the concept of a coincidence and in the arbitrary setting of significance levels, we can sometimes argue with any one who maintains a rival thesis to ours. But detailed dialogues are often tedious. If we are arguing with a sceptic, we may need to go over every step, countering each objection he makes and forcing him to put his own cards on the table and state what his alternative, null, hypothesis is. But if, as often, we are addressing ourselves to reasonable men, we can afford to omit some steps; and if, moreover, time is short, as it usually is, we need to economize time and argument. Often the price of this will be a lack of rigour and greater room for error: but often it will be a price worth paying. Provided we keep it in mind that our arguments are not watertight and that our conclusions should be tentative, it is often wiser to reach the best decision we can quickly, than to maintain a prolonged academic suspense of judgement. The best available view is better than no view, provided we are prepared to replace it by a better one still, should it become available. Therefore we turn from considering the tactics of the in-fighting when arguing with a sceptic, to devising strategies to enable us to advance as fast as possible on a broad front.

Instead of inviting our sceptical hearer to put forward an alternative hypothesis, we may invite our non-sceptical hearers to "see" that a whole class of alternative hypotheses are going to be ruled out—according to some level of significance—by the available evidence. If we toss a coin a thousand times and have 530 heads, although we cannot say with any show of reason that the probability of a head is exactly 0·53, we *can* say that *any* hypothesis assigning a probability of more than 0·58 or less than 0·48 could have yielded the results we have only by a one-in-several-hundreds coincidence. Therefore, granted such a level of significance, we can say that the probability of the coin coming down heads must lie between 0·48 and 0·58; or compendiously 0·53 ± 0·05.

It is dangerously easy to misconstrue this argument as a straightforward probabilistic argument, that if the frequency of heads is 0·53 then there is a 99% probability that the probability of a toss coming down heads is 0·53 ± 0·05, regarding it as an analogue to the valid argument that if the probability of a toss coming down heads is 0·53, there is a 99% probability that the frequency of heads in a thousand tosses will be 0·53 ± 0.05, that is, that there will be between 480 and 580 heads altogether. The latter argument is expressed graphically by the simple Binomial Distribution, in figure 8(i): the former needs the more complicated pattern of figure 8(ii) (see next page).

In figure 8(ii) we consider a whole class of (slightly skew) Binomial Distributions, and see which ones could yield 530 heads only by a coincidence greater than our chosen level of significance. These ones the reasonable man will dismiss, unless he has some further, special reason for supposing one of them to hold. For if he maintained that one of them did hold, then he would have to dismiss the actual frequency as a 1 in 20, 1 in 100, or a 1 in 1,000 coincidence. And this he will not do without good reason.

We need to be careful in what we say. We do not say that it is 95% (or 99% or 99·9%) probable that the real probability lies within the limits thus determined, but only that *if* it does not, something rather/highly/very highly improbable (*i.e.* 5% 1%, 0·1%) has happened. It is *modus tollendo tollens*. But *modus tollendo tollens* is a rule of two-valued propositional calculus, not of continuous-valued probability calculus. It is only by assimilating improbability to falsehood that we can use the rule at all. Having done this, we can proceed formally and reject as false any hypothesis outside the limits 0·53 ± 0·05. But then we are *rejecting* them as *false*, not merely assigning to them a low probability. We cannot by Bernoulli's Theorem assign any positive number as the probability that the 0·6 hypothesis is the wrong one, or that the (0·53 ± 0·05) is the right one, although, unfortunately for clear thinking, we can use the words 'probably' or 'probable' in their non-numerical sense simply to hedge and warn our hearer that our conclusion is not cast-iron, but depends upon a significance level whose exact value cannot be established by argument

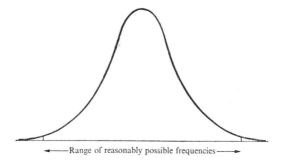

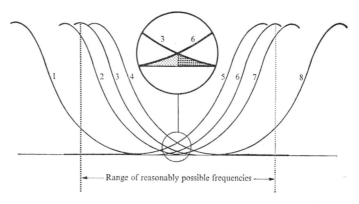

Fig. 8 (i). The Binomial Distribution. For a given probability of heads and a given number of tosses, it gives the probabilities of different frequencies actually occurring: if also a level of significance is given (marked by vertical lines), it gives range of reasonably possible frequencies, and of those possible only by a remarkable coincidence.

Fig. 8 (ii). For a given frequency in a given number of tosses, and a given level of significance, it gives probability of heads (marked in heavy dots) at or below which the given frequency could have occurred only by a remarkable coincidence; and probability of heads (marked in light dots) at or above which the given frequency could have occurred only by a remarkable coincidence; and hence the range of probabilities which could reasonably possibly (*i.e.* without a remarkable coincidence) have yielded the given frequency.

alone. In this sense it is only probable that the (0·53 ± 0·05) hypothesis is correct: but there is no mathematical probability less than 1 which we can assign to it on the basis of our observations. All we can say is that unless the (0·53 ± 0·05) hypothesis is correct, the observed results must be due to a coincidence implausibly improbable.

We want to say more; and some thinkers have attempted to develop Confirmation theory in order to be able to assign a numerical measure to the degree of support that a particular hypothesis obtains from a particular set of results. It is not a hopeless endeavour, though if it were to succeed the resulting theory would, for the reasons given earlier,† have to be radically unlike the calculus of probabilities. But I doubt if any useful way will emerge of quantifying the inverse argument based on Bernoulli's Theorem, just because the crux of the argument is the rejection of coincidences. We must not be too brusque with coincidences. Coincidences are not just improbable events to which one numerical magnitude can be assigned. As we have seen, the same event may be a remarkable coincidence under one description and quite unremarkable under another. Although we can give some guidance in selecting the right description, we cannot give a complete set of rules for deciding which description is the right one; for it depends on the alternative explanations we have to consider. And whether we should reject a coincidence as being too improbable, or accept it, notwithstanding the mathematical improbability assigned to it, depends not only on the degree of improbability, but on the relative merits of the two rival explanations under consideration, and these depend on other factors besides the actual results we have, other factors which may not be susceptible of numerical measurement. Confirmation theory is in danger of ignoring these qualifications and becoming a recommendation to assume that coincidences beyond a certain point be no longer believed in; and the fact that we have conventions about levels of significance encourages the unwary to go through life always rejecting as false anything whose probability falls below 5%. But it would be self-contradictory, as well as unwise in practice, to adopt such a rule. We often need to calculate with probabilities far smaller than any significance level ever proposed. In sub-atomic physic 10^{-10} is non-negligible. And always the decision whether to explain away a particular set of results as a

† In Ch. II, p. 12.

coincidence or not depends not only on calculations of probability but on our general notions of what types of explanation will wash. Coincidence is too much of an open-ended concept to be readily incorporated in a cut-and-dried calculus of confirmation, and I doubt whether the inverse argument based on Bernoulli's Theorem will ever be satisfactorily formalised.

The application of Bernoulli's Theorem is of central importance because it enables us to argue backwards from frequencies to probabilities, and thus assign numerical magnitudes to propositional functions with a fair degree of precision. If the argument of Chapter V is accepted, it provides a link, although tenuous and elastic, between probabilities and truth, so that sooner or later, although not at any generally established level of improbability, we can reject a hypothesis not merely as a strategic manoeuvre but as being actually false. It gives a logical justification for the common-sense practice of ascribing probabilities on the basis of frequencies. Nevertheless it has grave disadvantages too. Although the existence of significance levels is, as we have shown in Chapter V, not arbitrary, the exact choice of levels is. Moreover we can only dub hypotheses true or false and not assign any probability-value in their turn to them. And finally, the concept of coincidence needs firm handling, or absurdity will ensue. None of these arguments is decisive: but they are enough to lead statisticians to examine another, more probabilistic approach.

Bayes' Theorems are consequences of the unrestricted Conjunction Rule which gives two different expressions for $\mathrm{Prob}[HF(g)]$, namely

$$\mathrm{Prob}[H(g)] \times \mathrm{Prob}[F(gh)]$$

and $\qquad \mathrm{Prob}[F(g)] \times \mathrm{Prob}[H(gf)].$†

Equating these two expressions, and dividing by $\mathrm{Prob}[F(g)]$, we obtain Bayes' Rule

$$\mathrm{Prob}[H(gf)] = \mathrm{Prob}[F(gh)] \times \frac{\mathrm{Prob}[H(g)]}{\mathrm{Prob}[F(g)]}.$$

It is evident that Bayes' Rule will be likely to enable us to construct inverse arguments in probability: for, within a general

† See above, Ch. IV, pp. 66–7, and p. 60 n.2.

universe of discourse of G-type things, it tells us how to calculate the probability of an f being H given the probability of an h being F, together with a couple of other probabilities.

Sir Ronald Fisher gives the following example from genetics.†

In Mendelian theory there are black mice of two genetic kinds. Some, known as homozygotes (BB), when mated with brown yield exclusively black offspring; others, known as heterozygotes (Bb), while themselves also black, are expected to yield half black and half brown. The expectation from a mating between two heterozygotes is 1 homozygous black, to 2 heterozygotes, to 1 brown. A black mouse from such a mating has thus, prior to any test-mating in which it may be used, a known probability of $\frac{1}{3}$ of being homozygous, and of $\frac{2}{3}$ of being heterozygous. If, therefore, on testing with a brown mate it yields seven offspring, all being black, we have a situation perfectly analogous to that set out by Bayes in his proposition, and can develop the counterpart of his argument, as follows:

The prior chance of the mouse being homozygous is $\frac{1}{3}$; if it is homozygous the probability that the young shall be all black is unity; hence the probability of the compound event of a homozygote producing the test litter is the product of the two numbers, or $\frac{1}{3}$.

Similarly, the prior chance of it being heterozygous is $\frac{2}{3}$; if heterozygous the probability that the young shall be all black is $\frac{1}{2^7}$, or $\frac{1}{128}$; hence the probability of the compound event is the product, $\frac{1}{192}$.

But, one of these compound events has occurred; hence the probability after testing that the mouse tested is homozygous is

$$\frac{1}{3} \div \left(\frac{1}{3} + \frac{1}{192} \right) = \frac{64}{65},$$

and the probability that it is heterozygous is

$$\frac{1}{192} \div \left(\frac{1}{3} + \frac{1}{192} \right) = \frac{1}{65}.$$

If, therefore, the experimenter knows that the animal under test is the offspring of two heterozygotes, as would be the case if both parents were known to be black, and a parent of each were known to be brown, or, if, both being black, the parents were known to have produced at least one brown offspring, cogent knowledge *a priori* would have been available, and the method of Bayes could properly be applied.

† R. A. Fisher, *Statistical Methods and Scientific Inference*, Edinburgh, 2nd ed., 1959, pp. 18–20.

The strategy of the argument can best be conveyed diagrammatically:

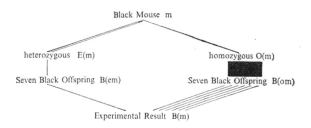

Black Mouse m

heterozygous E(m)   homozygous O(m)

Seven Black Offspring B(em)   Seven Black Offspring B(om)

Experimental Result B(m)

We start at the top with the experimental set-up—a black mouse of known parentage—we then plot the two *routes* by which this set up could yield the actual experimental result, and consider how probable it is that one route rather than the other is the one in question. At the first stage, the way down is much easier—twice as easy—if we suppose the mouse (m) to be heterozygous [$E(m)$, hEterozygous] rather than homozygous [$O(m)$, hOmozygous]. But the descent from being a heterozygous mouse (em) to having Seven Black Offspring ($B(em)$) is very tenuous and difficult—only a probability of $\frac{1}{128}$—whereas the descent from being a homozygous mouse (om) to having Seven Black Offspring ($B(om)$) is dead easy, being 100% sure. Hence it is relatively much easier to reach the actual experimental result by the right hand path—that is, on the assumption that the mouse is homozygous—than by the left hand path—on the assumption that the mouse is heterozygous. To be exact, the ratio of the probabilities is $\frac{1}{3} \times 1 : \frac{2}{3} \times \frac{1}{128}$, *i.e.* $1 : \frac{1}{64}$. Hence, given the actual experimental result of this black mouse having seven black offspring, it is 64 to 1 that he was a homozygous one.

We can put the argument formally in our symbolism thus: the mouse of known parentage can be referred to by the individual variable m. The propositional function that any such mouse be homozygous can be expressed by $O(m)$. The propositional function that any such mouse be heterozygous can be expressed by $E(m)$. The propositional function that a homozygous mouse, *om*, have seven black offspring can be expressed by $B(om)$. The propositional function that a heterozygous mouse, *em*, have seven black off-spring can be expressed by $B(em)$.

$$\text{Prob}[O(m)] = \frac{1}{3}, \quad \text{Prob}[B(om)] = 1, \quad \therefore \; \text{Prob}[BO(m)] = \frac{1}{3}$$

$$\text{Prob}[E(m)] = \frac{2}{3}, \quad \text{Prob}[B(em)] = \frac{1}{128}, \quad \therefore \; \text{Prob}[BE(m)] = \frac{1}{192}.$$

We want $\text{Prob}[O(bm)]$.
By Bayes' Rule,

$$\text{Prob}[O(bm)] = \text{Prob}[O(m)] \times \frac{\text{Prob}[B(om)]}{\text{Prob}[B(m)]}.$$

Now $\text{Prob}[O(m)]$ and $\text{Prob}[B(om)]$ are given.

$$\text{Prob}[B(m)] = \text{Prob}[BO(m) \lor BE(m)]$$
$$= \text{Prob}[BO(m)] + \text{Prob}[BE(m)];$$

hence the calculations. So, if we know the parentage of the mouse, and the corresponding probabilities of its being homo- or hetero-zygous, we can calculate exactly the probability of such-a-mouse-with-seven-black-offspring being homo- or hetero-zygous.

It is tempting to try to extend this argument, so that we can argue from observed frequencies to supposed probabilities with-out having to use any level-of-significance argument. Instead of the black mice of Fisher's example we consider *possible hypotheses*. Instead of this black mouse (the actual one used in the breeding experiment), we consider the true hypothesis. And to the results of the breeding experiment correspond the observed frequency. Thus if we toss a coin, the probability of its coming down heads could be anything between 0 and 1. The possible hypotheses are the non-denumerable set in which the probability of its coming down heads takes every value between 0 and 1. IF—and it is a very big if indeed—we knew the probability-distribution of these values, as we do know, in Fisher's example, the probability of this black mouse being homozygous or heterozygous, then Bayes' Rule would apply; if, for example, the probability-distribution was uniform, and there was $\frac{1}{2}$ a chance of its being between 0 and $\frac{1}{2}$, a $\frac{1}{4}$ of a chance of its being between $\frac{1}{2}$ and $\frac{3}{4}$, and, generally, δx of its being between x and $x + \delta x$: then a straight-forward Bayesian argument would go through, and we could argue about hypotheses as validly as we do about mice, with the experimental results corresponding to the mouse's having seven black offspring, and the hypothesis corresponding to the propositional function that the mouse is homo- (or hetero-)zygous. But there is a difficulty. In the homo-/hetero- case there are two "hypotheses" available, and there is given at the outset the probability of the propositional function that each is true. The hypotheses are not propositions, but propositional functions. In that particular case they range over mice-of-specified-parentage, but in general they will range over a universe of discourse of possible hypotheses. We start with a universe of possible hypotheses— *e.g.* that some parameter (often some probabilistic parameter) has some (often approximate) value; if we can at the outset assign a probability to the propositional function of a possible hypothesis actually being the particular one in question, then we can shift our universe of discourse, and calculate the probability of a-possible-hypothesis-which-has-in-fact-yielded-the-given-experimental-results being the particular one in question. The chief difficulty will, of course, be to assign *at the outset* any probabilities at all. But quite apart from this difficulty there is a conceptual one, surmountable but sufficient to trap the unwary, in thinking of the probability of a possible hypothesis being a particular one. It is much more natural first to specify a particular hypothesis completely, and then to consider the probability of its being true. But no calculation then can be made on the basis of Bayes' Rule.

Supposing the conceptual difficulty mastered, can we make any assignment of probabilities at the outset? It has often been supposed that we can; and, in particular, that if we have no reason to make any other assignment, then we are entitled to make one on a basis of equiprobability. In our example, where we are given a coin, and toss it a thousand times and find it comes down 530 heads, many writers have argued thus: the hypothesis that the probability of the coin coming down heads—$C(a)$, let us call it— lies in any one given interval is itself at the outset as probable as that it lies in any other interval (of the same size);

i.e.
$$\text{Prob}[\alpha \leqslant \text{Prob}[C(a)] \leqslant \alpha + \delta\alpha] = \delta\alpha.$$
The probability of a coin for which
$$\text{Prob}[\alpha \leqslant \text{Prob}[C(a)] \leqslant \alpha + \delta\alpha] = \delta\alpha$$

coming down 530 times heads in 1,000 tosses can be calculated
for any α and any sufficiently small $\delta\alpha$. Hence, integrating $\delta\alpha$
from 0 to 1, we can calculate the total probability of our getting
530 heads in 1,000 tosses, taking all hypotheses about the value of
Prob[$C(a)$] together. Hence, using Bayes' Rule, we can calculate
the probability of $\alpha \leqslant$ Prob[$C(a)$] $\leqslant \alpha + \delta\alpha$, given that there
were 530 heads in 1000 tosses. Roughly, if Hyp$_\alpha^{\delta\alpha}(g)$, or $H(g)$
for short, expresses the propositional function that a hypothesis—
g—assigns to Prob[$C(a)$] a value between α and $\alpha + \delta\alpha$, and gf is
a hypothesis which yielded the observed results (530 heads in
1000 tosses), then

$$\text{Prob}[H(gf)]-i.e.\ \text{Prob}[\text{Hyp}_\alpha^{\delta\alpha}(gf)]-=\frac{\text{Prob}[H(g)]\times\text{Prob}[F(gh)]}{\text{Prob}[F(g)]}$$

and it is clear that the right-hand side is proportional to
Prob[$F(gh)$].

This is Bayes' First Theorem, sometimes known as The Inverse
Principle of Maximum Likelihood. It states:

If a propositional function has come true on m out of n occasions that
are similar and independent of one another, then the most probable
value for the probability of the propositional function is around m/n,
provided that all hypotheses assigning a probability-value to the propositional function are each, apart from the information that the propositional function has come true on m out of n occasions, as probable
as any other hypotheses.

Again, a diagram makes the argument easier to grasp. In this case
we distinguish only three levels. At the top we have the universe

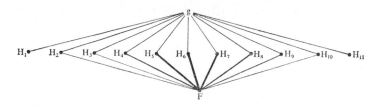

of possible hypotheses, g. At the second level we have the particular hypotheses, $H_1, H_2, \ldots, H_n$. In the diagram and in the formal
proof I shall assume that there is only a definite finite number of
possible hypotheses each having the same, non-infinitesimal, probability. When we have introduced the concept of probability-

density in Chapter X, those readers familiar with integrals will have no difficulty in generalising the proof to cover the continuous case: and those readers who always shut their eyes when they see an integral sign may be reassured to know that with Bayes' Theorems all the philosophical issues arise just as much with the discrete as with the continuous case. At the bottom we have the actual experimental result, F. The lines from h to H_1, $H_2, \ldots, H_n$, are all equally thick, to represent the fact that they all have the same initial probability: but the lines between the middle and bottom levels are not equally thick, because the probability of getting the experimental result F from gh_1—i.e. if $H_1(g)$ is true—is not the same as getting it from gh_2—i.e. if $H_2(g)$ is true. This is because the different hypotheses gh_1, gh_2, $\ldots$, etc. are themselves probabilistic hypotheses—that is, they are hypotheses that the real probability, in terms of which the result F is to be explained, has a certain value. For example, if we could assume that the probability of a coin coming down heads was either 0 or 0·1 or 0·2 or 0·3 or 0·4 or 0·5 or 0·6 or 0·7 or 0·8 or 0·9 or 1·0, and that the probability of each of these hypotheses was initially equal, then there would be a $\frac{1}{11}$ probability that $H_1(g)$ was true, and the probability of heads was 0, but on this hypothesis gh_1, there would be no chance of our getting, say, 530 heads in a thousand tosses; i.e. $F(gh_1) = 0$. On the assumption that the probability was 0·1, gh_2, i.e. the assumption that $H_2(g)$ was true, there would be a small probability—viz. $^{1,000}C_{530} \times (0\cdot1)^{530} \times (1 - 0\cdot1)^{470}$—of getting 530 heads; i.e. $F(gh_2) = {}^{1,000}C_{530} \times (0\cdot1)^{530} \times (0\cdot9)^{470}$.

Similarly if $H_3(g)$ were true, and the probability was 0·2, there would be another small, but less small, probability of our getting 530 heads; for

$$F(gh_3) = {}^{1,000}C_{530} \times (0\cdot2)^{530} \times (0\cdot8)^{470}.$$

And so we continue, calculating $F(gh)$ for each value of h. It seems reasonable, although we have not proved it yet,† that the probability will be greatest for gh_6—viz. the assumption that the probability of getting heads in a toss is 0·5: for we have proved the converse, that for a given probability the most probable frequency is the nearest possible one to the given probability. We represent this on the diagram by drawing a much thicker line

† We shall, on p. 154.

from gh_6 to F, and a somewhat thicker one from gh_7 to F, *etc.*, than from gh_2, gh_{10}, *etc.* We then argue that since the first stage of the descent is equally easy whichever path we take, the relative ease of the various *routes* the whole way from top to bottom must be proportional to that of the second half.

The formal proof is straightforward, though somewhat heavy. For the sake of clarity we need first, rather redundantly, to specify the propositional function under consideration to which we shall be assigning a probability: in our example, the propositional function that a particular coin when tossed will come down heads; we have called it $C(a)$. We then have another propositional function that the-probability-of-$C(a)$ is such-and-such; let us call this second-order propositional function $H(g)$, and let H_1, $H_2, \ldots, H_m, \ldots, H_r$ be the finite set of mutually exclusive possible hypotheses about the correct value of $C(a)$: that is, the disjunction $H_1(g) \lor H_2(g) \lor \cdots \lor H_m(g) \lor \cdots \lor H_r(g)$ is true. Let $F(g)$ be the propositional function that, with some assignment or other of probability to $C(a)$, in a sequence of n trials $C(a)$ comes true in m of them. Let $H_l(g)$ be the hypothesis which assigns to $C(a)$ the value nearest m/n.

$F(g)$ may be true not only when $H_l(g)$ is true, but when $H_1(g)$, $H_2(g)$, $\ldots$, $H_{l-1}(g)$, $H_{l+1}(g)$, $\ldots$, $H_r(g)$ are true. So

$$F(g) = FH_1(g) \lor FH_2(g) \lor \cdots \lor FH_r(g).$$

$\therefore$ Prob$[F(g)$

$\qquad = \text{Prob}[FH_1(g)] + \text{Prob}[FH_2(g)] + \cdots + \text{Prob}[FH_r(g)]$

by the Exclusive Disjunction Law, since $H_1(g)$, $H_2(g)$, $\ldots$, $H_r(g)$ are all mutually exclusive hypotheses. Each term on the right-hand side can be expressed, in virtue of the unrestricted Conjunction Rule as a product of two probabilities. Thus

$$\text{Prob}[FH_1(g)] = \text{Prob}[H_1(g)] \times \text{Prob}[F(gh_1)]$$

and $\qquad \text{Prob}[FH_2(g)] = \text{Prob}[H_2(g)] \times \text{Prob}[F(gh_2)]$

$\cdot \quad \cdot \quad \cdot \quad \cdot \quad \cdot \quad \cdot \quad \cdot \quad \cdot \quad \cdot \quad \cdot \quad \cdot \quad \cdot \quad$ *etc.*

The first factor on the right-hand side of each product is the same by the assumption of equiprobability. We can therefore express the whole

$$\text{Prob}[F(g)] = \text{Prob}[H_1(g)] \times \{\text{Prob}[F(gh_1)]$$
$$+ \text{Prob}[F(gh_2)] + \cdots + \text{Prob}[F(gh_r)]\}.$$

We can now use Bayes' Rule to calculate $\text{Prob}[H_i(gf)]$ which is what we want to know.

$\text{Prob}[H_i(gf)]$

$$= \frac{\text{Prob}[H_i(g)] \times \text{Prob}[F(gh_i)]}{\text{Prob}[F(g)]}$$

$$= \frac{\text{Prob}[H_i(g)] \times \text{Prob}[F(gh_i)]}{\text{Prob}[H_i(g)]\{\text{Prob}[F(gh_1)]+\text{Prob}[F(gh_2)]+ \cdots +\text{Prob}[F(gh_r)]\}}$$

$$= \frac{\text{Prob}[F(gh_i)]}{\text{Prob}[F(gh_1)] + \text{Prob}[F(gh_2)] + \cdots + \text{Prob}[F(gh_r)]}$$

since $\text{Prob}[H_i(g)] = \text{Prob}[H_1(g)] = \text{Prob}[H_2(g)]$, etc. The denominator is the same in every case; $\text{Prob}[H_i(gf)]$ is therefore proportional to $\text{Prob}[F(gh_i)]$.

It only remains to prove that $\text{Prob}[F(gh_i)]$ is the greatest of $\text{Prob}[F(gh_1)]$, $\text{Prob}[F(gh_2)]$, ..., $\text{Prob}[F(gh_r)]$, as we conjectured. If $H_i(g)$ assigns the probability value α to $C(a)$, then $\text{Prob}[F(gh_i)] = {}^nC_m\alpha^m(1 - \alpha)^{n-m}$ which we write ${}^nT_m$ for short. It is a purely mathematical exercise to determine the value of α for which ${}^nT_m$ is a maximum. Suppose, first, that α is continuous —this is not to suppose that we are dealing with continuous probability-densities, but only to solve a purely mathematical problem—, then the maximum value of ${}^nT_m$ for different α's is obtained by differentiating with respect to α, and solving for when

$$\frac{\mathrm{d}\,{}^nT_m}{\mathrm{d}\alpha} = 0,$$

$$\frac{\mathrm{d}\,{}^nT_m}{\mathrm{d}\alpha} = \frac{\mathrm{d}}{\mathrm{d}\alpha}\,{}^nC_m\alpha^m(1 - \alpha)^{n-m}$$

$$= {}^nC_m[m\alpha^{m-1}(1 - \alpha)^{n-m} - (n - m)\alpha^m(1 - \alpha)^{n-m-1}]$$

$$= 0 \quad \text{when} \quad m(1 - \alpha) - (n - m)\alpha = 0,$$

i.e. when $\alpha = m/n$. Even if we have only discrete hypotheses, whatever discrete set of values they assign to α, ${}^nT_m$ will be less for those values than for $\alpha = m/n$. Hence $\text{Prob}[F(g_i)]$ will be larger than any other $\text{Prob}[F(g_r)]$, and so likewise $\text{Prob}[H_i(fg)]$; which proves Bayes' First Theorem, The Inverse Principle of Maximum Likelihood. The result is intuitively acceptable: but the initial assumption of equiprobability is open to grave objection. We do not know that the probability is evenly distributed, and we have no warrant for assuming it. And therefore, in the view of many critics, no Bayesian argument is possible.

Not all equiprobability arguments are absurd. As we have seen,† there can be a rational presumption of equiprobability, although one always liable to be defeated by actual evidence. The difficulty is not that all assumptions of equiprobability must be wrong, but that the ones actually made are unjustifiable. In the cases of dice, roulette wheels and cards we argued for equiprobability on the grounds that any other assignment of probabilities would need explaining, and that we had reason to suppose that no such explanation could be given. No similar repugnance to non-equal probabilities can be sustained when we assign them to possible hypotheses. We should not have to revise our laws of nature if it turned out to be, at the outset, more probable that the probability of a coin's coming down heads was about $\frac{1}{2}$ than about $\frac{1}{10}$ or $\frac{9}{10}$: rather, if anything, the contrary. Not only, in this case, is there no reason for thinking the initial probabilities equal, but there is some reason for thinking them not. Only a very democratic Platonist, who believed that all possible theories existed in a Platonic heaven, and that all were endowed there with equal opportunity of being right, could really use the Bayesian argument as I have stated it.

Nevertheless, something may be saved from the wreck. We may not know what the "initial" ("prior", "a priori") probabilities are exactly, but may have a fair idea of the limits within which they lie, or the general shape of the distribution curve—often we have reason to believe that the distribution is a Normal Distribution.‡ Even if we cannot say anything very positive, we may be able to rule out, or at least discount, some extremely singular cases. For such singularities need explanation. If, for instance, the probability of a coin's coming down heads was almost certainly less than 0·001, then the Bayesian argument would not only be invalid but would lead to quite erroneous conclusions. But it would need a lot of explaining. As sometimes with randomness, so with non-singularity: it may be presumed, although the presumption may subsequently be rebutted. But in the absence of counter-evidence, the presumption is a rational one.

We can make use of this weaker presumption because the actual

† In Chapter VI.   ‡ See below, Ch. X, pp. 182–6.

assignment of "initial" probabilities does not carry as much weight in the final assessment as the observed results. The mouse with seven black offspring had a very high probability of being homozygous, although initially it had only $\frac{1}{3}$. Even if it had been initially only $\frac{1}{4}$, this would have been more than counterbalanced by the birth of an eighth black offspring; while if it had been initially $\frac{2}{3}$, we should have still needed five black offspring to reach the same final probability of $\frac{64}{65}$. Whatever, within wide limits, the "initial" probabilities, the final ones will be, in the long run, more or less the same, and will mirror only the observed results, not the initial preconceptions. It thus seems to be hair-splitting to argue whether "initial" probabilities are equal or not when it does not really matter either way. Only if we have very scanty evidence do our initial assumptions matter much. Otherwise, even if they are wrong, their error will be drowned in the welter of statistical fact, and will not seriously affect our conclusions. Therefore we need not consider what exactly the "initial" probabilities are. Provided they lie within reasonable limits, and provided we have a reasonable amount of evidence, we can argue confidently and fairly accurately.

This is the burden of Bayes' Second Theorem, which is sometimes known as the Inverse Law of Great Numbers. It states:

If the relative frequency with which a propositional function comes true is m/n, on a number, n, of occasions that are similar and independent of one another, then the probability of the hypothesis assigning to the propositional function the probability-value m/n approaches as a limit the maximum value 1, as n increases indefinitely, provided that the probability of that hypothesis, apart from the information that the relative frequency is m/n, does not have a probability-value 0.

The proof is similar to that of Bayes' First Theorem, except that instead of the assumption of equiprobability we have $\text{Prob}[F(gh_r)] \to 0$ as $n \to \infty$ for $r \neq l$. Therefore, if we have only a finite number of hypotheses, since their total probability is 1, $\text{Prob}[F(gh_l)] \to 1$.† Therefore by Bayes' Rule,

$$\text{Prob}[H_r(gf)] = \frac{\text{Prob}[F(gh_r)] \times \text{Prob}[H_r(g)]}{\text{Prob}[F(g)]} \to 0 \text{ as } n \to \infty$$

for $r \neq n$, since $\text{Prob}[F(g)] \neq 0$.

† There is a hole in the argument here, which blocks itself when we generalise to the continuous case.

And

$$\text{Prob}[H_i(gf)] = \frac{\text{Prob}[F(gh_i)] \times \text{Prob}[H_i(g)]}{\text{Prob}[F(g)]}.$$

and this does not equal 0, provided $\text{Prob}[H_i(g)]$ does not—which is given *ex hypothesi*. Then, since one of the hypotheses, $H_1(gf)$, $H_2(gf)$, ..., $H_r(gf)$ must be true,

$$\text{Prob}[H_1(gh)] + \text{Prob}[H_2(gh)] + \cdots + \text{Prob}[H_r(gh)] = 1,$$

whence $\text{Prob}[H_i(gh)] \to 1$.

But are we entitled thus to limit the possible probability-values of the possible hypotheses at the outset? We answer by arguing that at least some initial assignments of probabilities would, in the absence of explanation, be implausible. If there were a uniform, high probability for assignments of probabilities between 0 and 0·1, and zero elsewhere, we should indeed want to know why. And even in advance of an explanation's being offered, we can almost rule it out as being incompatible with our general theories about coins and coin-tossing. We can, in theories about macroscopic phenomena, assume continuous, perhaps even analytic, distribution functions for parameters—often, indeed, Normal ones: this on the same grounds as we assume equiprobabilities for dice, roulette-wheels and cards. We always could be wrong. But if we were wrong, the explanation of why we were wrong would fit so ill with other well-tried theories, that we should have to scrap them: and this we are reasonably reluctant to do. The sceptic who merely queries the old, equiprobability assumptions of Bayesian arguments can easily score points: but the sceptic who claims that the results reached are actually wrong has to make Bayesian counter-assumptions which are more wildly implausible than any that the simple-minded Bayesian ever makes.

Any neo-Bayesian argument should be cast into dialectical form; one should invite one's hearer to put forward assumptions strong enough to vitiate one's conclusions, and then argue from there. It is a fair challenge. "Initial" probabilities can be used to counter a straightforward level-of-significance argument. In the example of the mice, if we had been operating with a 1% level of significance, we should, in the absence of other information, have concluded that the mouse in question was homozygous, since there is only a 1 in 128 chance of a heterozygous mouse having seven black offspring by a brown mate. But with the additional information about its parentage, we calculated that there was a 1 in 65 chance of its being heterozygous, and this, at the 1% level, is non-negligible. We might therefore keep in mind the possibility that it was heterozygous, and had, by a remarkable coincidence, produced seven black offspring. Certainly, if we had known that the chances were one million to one against any black mouse being homozygous, then we should believe that his offspring were black by coincidence rather than that he was by a much longer coincidence homozygous. However, although "initial" probabilities can bear in the opposite direction to the observed frequency, their influence is not very great. In spite of a 2 to 1 chance in favour of being heterozygous at the outset of the experiment, the final verdict was 64 to 1 against. If we had been using either the 5% or the 0·1% level of significance, our conclusion would have been unaltered by the "initial" probabilities. In general, the "initial" probabilities have much less effect than any observed frequencies that we might regard as statistically significant. Only if the "initial" probabilities are extreme, will they seriously counter the effect of statistically significant observed frequencies: and if they are extreme, some explanation will be called for. One can therefore make the Bayesian argument work after a fashion by requiring one's listener to posit—within wide limits—the "initial" probabilities he thinks are necessary. If they are at all ordinary, the Bayesian argument will go through, even though the calculation may be messier (and rather more observed frequencies may be required) than in the equiprobable case: if they are quite extraordinary, then we ask the reason why. And, if it is a good reason, our observations will have been largely unnecessary. If we know enough to doubt the Bayesian argument's applicability, we do not need it: if we do not, then the argument is fair enough. It is not watertight—how could any argument founded on ignorance be that? But it is not totally fallacious. It may turn out in a particular case to be wrong: there may be strong, though unsuspected, reasons why the true hypothesis is very probably some singular one. In that case, we shall discover our error in the fulness of time, as more evidence accumulates. But we need not be paralysed by this possibility. All probabilistic arguments are only tentative, anyhow. Although we may be wrong, it is reasonable to pose this dilemma: either the initial probabilities are extraordinarily singular or they are not. If they are, there must be some reason for it, and this reason will itself give us a fair indication of what the values are without more ado. If they are not, then a Bayesian argument will determine, not exactly but within useful limits, what the actual values are.

It is difficult to say what the limits are. It depends on the case in question. What is reasonable in one set of circumstances may be quite unreasonable in another, and *vice versa*. All we can do in general is to consider what effect different limits would have on the conclusions we reach. If the initial probability of the black mouse's being homozygous had been $\frac{1}{9}$ instead of $\frac{1}{3}$,

$$\text{Prob}[BO(m)] = \frac{1}{9}, \text{ and } \text{Prob}[BE(m)] = \frac{8}{9} \times \frac{1}{128} = \frac{1}{9} \times \frac{1}{16} = \frac{1}{144}.$$

So

$$\text{Prob}[B(m)] = \frac{1}{9}\left(1 + \frac{1}{16}\right) = \frac{1}{9} \times \frac{17}{16},$$

$$\text{Prob}[O(bm)] = \frac{16}{17}, \text{ and } \text{Prob}[E(bm)] = \frac{1}{17}.$$

If the initial probability had been $\frac{8}{9}$ instead of $\frac{1}{3}$,

$$\text{Prob}[BO(m)] = \frac{8}{9}, \text{ and } \text{Prob}[BE(m)] = \frac{1}{9} \times \frac{1}{128},$$

so

$$\text{Prob}[B(m)] = \frac{1}{9}\left(8 + \frac{1}{128}\right) = \frac{1}{9} \times \frac{1,025}{128}$$

and

$$\text{Prob}[O(bm)] = \frac{8}{9} \div \left(\frac{1}{9} \times \frac{1,025}{128}\right) = \frac{1,024}{1,025}.$$

In this case the change of initial probability from $\frac{1}{9}$ to $\frac{8}{9}$ would have shifted the final probability from 94% to 99·9%. Roughly, it would represent a shift from a 5% level of significance to a 0·1% one. If we thought it reasonable to assign to the hypothesis an initial probability within the interval (0·11, 0·89), then we could pin down the final probability to between 0·94 and 0·999. Alternatively, if we specified an acceptable range of variation for the final probability, we could calculate the interval within which the initial probability needs to lie in order to yield the desired conclusion.

Bayes' Second Theorem is thus become a blunt instrument. We cannot use it to calculate precise probabilities except in the unusual case where we are able to feed in precise initial probabilities. Where these are not given, we can at best feed in a range within which, we have reason to suppose, the initial probabilities must lie, and from this we may be able to calculate another range, often a surprisingly small one, within which the final probability must lie. Like the inverted use of Bernoulli's Theorem, it is essentially inexact: but, unlike Bernoulli's Theorem, it can be used to calculate, although only approximately, any probabilities, and not only those in the neighbourhood of 0 and 1.

Besides being applied in particular cases, Bayes' Rule can be used to justify certain procedures or "strategies". Strategies are rules of thumb for making decisions on the basis of various—and often varying—experimental results. They are not—and are not alleged to be—very reliable in individual cases, but if adopted will in the long run of cases maximise some desirable results— *e.g.* getting the answer right—or minimise some undesirable one —*e.g.* making mistakes. The "Maximum Likelihood" procedure (so named because $\mathrm{Prob}[F(gh)]$ is called the likelihood of H) instructs the experimenter to choose that hypothesis which has the results actually obtained as its most probable result: *i.e.* that h for which $\mathrm{Prob}[F(gh)]$ is a maximum. For this, it will be shown, means that $\mathrm{Prob}[H(gf)]$ is also maximised.

As before, we are considering possible theories, g, and among them possible theories having yielded certain experimental results, F. We denote such theories by gf. We want our possible theory, g, to be that one, H, for which the likelihood is a maximum. Now, by Bayes' Rule, the likelihood,

$$\mathrm{Prob}[H(gf)] = \frac{\mathrm{Prob}[H(g)] \times \mathrm{Prob}[F(gh)]}{\mathrm{Prob}[F(g)]}. \tag{i}$$

But, as we have argued, if there are a lot of experimental results, $\mathrm{Prob}[H(g)]$ does not matter much, since $\mathrm{Prob}[F(gh)]$ varies fairly sharply with h: and $\mathrm{Prob}[F(g)]$ is independent of our particular choice of h. We therefore concentrate on the $\mathrm{Prob}[F(gh)]$, and re-express (i) as

$$\mathrm{Prob}[H(gf)] \propto \mathrm{Prob}[F(gh)]. \tag{ii}$$

Given any possible theory g, and a particular version of it, h

(usually in virtue of an adjustable parameter α; $H(g)$ is the theory we get when α takes a particular value), it is feasible to calculate $F(gh)$ (the probability that the theory, for the given value of α, would yield results F). We can then determine for which value of α—*i.e.* for which h—this probability is greatest. And then infer that this is the actual value of α in question.

The Maximum Likelihood procedure will be a reasonable one to adopt in cases where there is a fair amount of experimental evidence, for then $\text{Prob}[F(gh)]$ usually has a fairly sharp maximum as α takes on different values, and $\text{Prob}[H(g)]$ has relatively little bearing on the final result: and if at any stage it is still introducing any significant error, the error will tend to be reduced as further evidence is obtained. It is therefore a reasonable strategy to adopt, although necessarily not a perfect one, and not necessarily the best one. In particular, it is bound to yield absurd conclusions in the extreme case where we have only one instance: for example, if we toss a coin once and it comes down heads, the Maximum Likelihood procedure would have us conclude that the probability of its coming down heads is 1, since on that hypothesis the result obtained is more likely (in fact, certain) than on any other.

Inverse probabilities are thus neither so valid as Bayes and Laplace thought, nor so vitiated as their modern critics have maintained. We cannot calculate a formula which we can always apply to a set of figures in order to determine the probability that the real probability in question has a particular value—the critics are right. But in many cases we can estimate it, within limits, granted certain, reasonable, assumptions, with a fair show of reason—Bayes and Laplace were not too far wrong. They were wrong in converting an argument from silence into an explicit assumption of equiprobability, but not in allowing some force to the argument from silence.

Altogether, there are three basic types of argument which enable us to ascribe fairly precise probabilities to propositional functions with a reasonable degree of confidence: each has its merits, but none is altogether satisfactory. An assumption of equiprobability requires only very general considerations of the form we expect natural laws to take, but it is valid only in special —often artificial—circumstances, and is a defeasible assumption which is always liable to be rebutted. The application of Bernoulli's Theorem relies least on background assumptions and most on observations actually made, but even so requires some assumptions—similarity, independence. It involves some element of arbitrariness, and moreover cannot give a probabilistic answer whether any particular hypothesis about the correct probability value is to be accepted or rejected, but only a black-or-white, yes-or-no one. The application of Bayes' theorem is not inherently arbitrary and enables us to assign probabilities, and not only the values True and False, to hypotheses: but usually requires us either to make assumptions that are unjustified or to be content with conclusions that are imprecise. Nevertheless, all three approaches can be used, and in variant forms and different combinations constitute powerful statistical techniques.

IX

TALKING WITH STATISTICIANS

STATISTICS is an emotionally charged subject. Many people fear it and hate it. Some revel in it and worship it. Inside the discipline the Montagues engage in wordy battles with the Capulets, and sometimes even sue them in the courts for holding wrong opinions. Most statisticians despise most other statisticians, and all despise all non-statisticians, who would like to be able to return the compliment, but are not quite sure how to shrug off a "best fit". In an atmosphere of mutual suspicion and contempt, neither side is willing to learn what the other has to teach. Statisticians labour to answer questions no one would ever want to ask, or make mountains of figures bring forth a mouse which was already common knowledge and needed no further proof: while laymen continue to hug their cherished beliefs that they can smoke without hazard to themselves and drive drunkenly without hazard to others, dismissing all evidence to the contrary as damned lies or even worse. We need to talk to statisticians. They need to know what are the questions we should like them to answer, and what are the assumptions it is reasonable to make, and we need to get behind their bare *ipse dixit*, and discover exactly what the question is to which they think they have the answer, and roughly what assumptions they are making, and very roughly what their pattern of inference is, and what price we should have to pay in order reasonably to deny their conclusions. Ideally, the non-statistician should have some knowledge of statistics, and the statistician should have some knowledge of, and feeling for, the field in which he is exercising his statistical techniques. For statistics can answer only certain sorts of questions, and it is for the non-statistician to decide which of these, if any, he wants to ask. And often the statistician needs to be able to make assumptions of irrelevance, and to have a background knowledge of the sort of natural laws which may be involved. But conditions are seldom ideal. It is not simply the gulf between

Lord Snow's Two Cultures: rather, it is a separate division, occurring even in the sciences, between the ultra-numerate statistician, and the actual practitioner who, even if he can face mathematics, does not like too many numbers, and is always wanting to get through them to the general pattern behind, without wanting to face either the discipline or the drudgery that statistics demands.

For these and other reasons there is often a breakdown of communications across the "interface" between statistics and other subjects. Much work needs to be done, and in this brief chapter I shall not attempt to do justice to the variety and subtlety of statistical techniques, but only to give a few crude—and sometimes unfair—hints to the layman who has to talk to the statistician. They are no substitute for understanding statistics, but may, I hope, help him elicit from a statistician some measure of understanding of the particular arguments employed, and the weight to be attached to them, in particular cases. The first and most important step is for each party to try and frame as exactly as possible the question to which an answer is being sought or is being given. The second is to distinguish stages which are purely computational from those where some inference is being made. The third is to pick out where and in what way probabilistic assumptions are being introduced or probabilistic conclusions reached. And finally we should make the statistician's argument undergo a "destruction test", by considering what else, if the conclusion were to be false, would have to be false too.

Framing the questions is more difficult than it seems. There is a natural tendency to be too agreeable. The statistician is liable to accept the layman's formulation and to assure him that it is just this question that he can answer, and the layman feels a bit bemused by the statistician's formulation, and does not like to say that he suspects it of being a dull question to which he does not want an answer. Instead, each party agrees to the formulation the other offers, while continuing to interpret it in his own way. It is necessary in framing the questions, as often in other philosophical activities, not simply to use words which can mean what we want them to mean, but to choose ones which cannot carry any other meaning than the one we intend. In particular, one needs to make sure that the statistician is not using words in a technical sense, unknown to the layman: 'best fits' may be defined in a clear and exact way, but to assume then that best fits are best is to commit a naturalistic fallacy.

When the questions have been framed, they may turn out to be unanswerable or unaskable. The question the layman is putting to the statistician may be one that statistics cannot answer. Statistics involves numbers; numbers of instances, and information expressed by numbers. Because it deals only with many instances, statistics cannot do justice to any one, and cannot convey to us the unique combination of properties which gives the individual its individual essence: but because it gathers data only in numerical form, and only care, not judgement, is required to obtain numerical answers to questions, statistics seems free from the pervasive subjectivism that infects humane disciplines. Both are fair points: neither decisive. It is true and important that statistics are useless for some purposes, and that it is better to have a qualitative, though possibly subjective, assessment of the factors I am interested in than an exact quantitative measure of something else. If I want to know what an individual person is like, or the climate of opinion in a society, or the "tone" of a school, I shall do far better to consult a well-informed observer than to carry out a statistical survey with a sociological questionnaire: but if I want life-expectancies, I shall do better to stick to the tables than to consult the people concerned. Statistics provides the right way to handle aggregates of quantitative data, and is proof against some sorts of subjective bias. But it is wrong to distrust men's judgement so completely that we must always have recourse to statistics to avoid it. Irrelevance is as bad as bias. Counsel is confused if, instead of an admittedly fallible assessment of what we want to know, we are given authoritative answers to questions we have not asked. In recent years politicians and economists have talked much about "economic growth"; but in order to measure it precisely, they have had to disregard any factor, however relevant to our material wellbeing, which does not appear in some trading transaction; with the paradoxical consequence that I am better off if I have bought three cars, none of which go very well, than if I have one which is always reliable, and that the value of our Health Service is calculated according to its cost, and not our ensuing health.

We must not, however, dismiss quantitative evidence too readily. It may not be conclusive, but it does not follow that it is irrelevant. Often the question the statistician can answer, although not the actual question we were asking, has some bearing on it. We ask whether the children of today are healthier than those of thirty years ago. The statistician tells us that on average they are taller and heavier, age for age, and infant mortality rates have declined. These are telling indications. It is possible to question them—we might say that children now were overweight, and suffered from nervous debility as a result of too much television—but the onus of proof lies on us. The statistician has not completely answered our question, but the answers he has given to his own questions go a long way towards answering ours; and if we are not going to go the whole way, we need to have some fairly definite and convincing alternative hypothesis to explain why we should not.

Much of the work a statistician does is purely computational. He has records of all the heights and weights of school children, and wants to answer his own question whether the average height and average weight has increased. The *mean*, or average, usually denoted by the Greek letter μ, or by $\bar{x}$, $\bar{y}$, $\bar{z}$, etc., is the simplest and best understood of all statistical parameters. If we have a finite set (*e.g.* of people), each one of which has associated with it a finite magnitude (*e.g.* height), the mean is calculated by dividing the sum total of all the magnitudes by the number of members of the set.

$$\mu = \frac{x_1 + x_2 + \cdots + x_n}{n}.$$

The *standard deviation* is less well known but equally important. It is usually denoted by the Greek letter σ, and in this simple case defined by the equation

$$\sigma = \sqrt{\frac{(x_1 - \mu)^2 + (x_2 - \mu)^2 + \cdots + (x_n - \mu)^2}{n}}.$$

Readers familiar with elementary mechanics may find it helpful to compare the mean with the centre of mass and the standard deviation with the moments of inertia. In a manner of speaking the standard deviation tells us to what extent the average man is *unlike* the average man. Each of the terms $(x_1 - \mu)^2$, *etc.* measures how far the individual concerned diverges from the mean, without regard to sign, and thus the standard deviation provides a measure, as its name suggests, of the deviation from the mean. It tells us, roughly, whether nearly all children are very close to average height more or less, or whether there is a wide variation between taller and shorter children. Or, perhaps better, we should distinguish the 'average' of everyday speech from the statistician's 'mean', and say that the standard deviation gives us the range around the mean which we should regard as average. A person is of average height if his height is between $\mu - \sigma$ and $\mu + \sigma$, and tall if he is more than $\mu + \sigma$. The standard deviation is clearly of importance for clothes shops; and whenever any argument is based on a change of the mean over the years, it is pertinent to know how the change compares with the standard deviation. For simple distributions the mean and the standard deviation can be calculated fairly easily. For the Binomial Distribution† the mean is given, very much as we should expect, by the equation

$$\mu = n\alpha$$

and the standard deviation by the equation

$$\sigma = \sqrt{n\alpha(1 - \alpha)}.$$

The fact that the standard deviation increases only as the square root of n, whereas the mean increases directly with n, is of some importance. It underlies the proof of Bernoulli's Theorem, and is crucial in the generalisation from the Binomial to the Normal Distribution.‡

The mean and the standard deviation (and other statistical parameters which I shall not discuss) enable the statistician to distinguish for us the wood from the trees. From the confusing welter of information, the statistician extracts the essential information, and suppresses the irrelevant rest, in much the same way as in ascertaining the probability of a coin's coming down heads, we record only the proportion of heads and tails in a sequence of tosses, and not the order. It is natural for a layman to supect the statistician of suppressing important truths, but nearly always wrong. Although what is relevant depends on the question being asked, and we always may put forward an alternative hypothesis which would require the figures to be worked over again, in

† See above, Ch. V, pp. 74–9.
‡ See below, Ch. X, p. 181, and Appendix I, pp. 211–2.

practice the range of serious hypotheses is limited, and they all will have substantially the same canons of relevance. The statistician is likely to have extracted most of what could be relevant to any reasonable hypothesis, and the layman who asks for more is in danger of committing the Dover fallacy.

Rather than query the computations, the layman should attempt to identify, and if possible to understand, the inferences. In some cases, we cannot expect much enlightenment, but we can at least ask for a name. We may not be able to follow sequential analysis; a chi-square test may be Greek to us, and "Student"'s t-test something we have never mastered, and unless we are very brave we may not trust ourselves to argue about Fiducial Arguments or have much faith in the confidence limits within which arguments can be relied on. But to have a name may reduce our timidity a little, and may enable us to consult a rival statistician, who may opine that a Neyman–Pearson strategy, however expedient in the long run for testing lots of hypotheses, may not give us a "best" result in this individual case, in the ordinary sense of best. At the least, it will make our own statistician clear his mind about the arguments he is using, and the criticisms to which they are sometimes subjected.

The weak link in most statistical arguments is at the point where probabilistic assumptions are being introduced or probabilistic conclusions interpreted. Often, indeed, it is quite hard to discover where these points are, because the probability attaches to the *method* rather than the subject under investigation. In Fisher's example of the old lady tasting tea,† a randomising device is used to determine the order in which the cups of tea are served. This is enough to prevent the old lady spotting the correct order by reason of some adventitious circumstance such as the glint in the experimenter's eye, but does not rule out the possibility of her random guesses happening to coincide with the actual order used. But if the actual order to be used is determined by a randomising device like a pack of cards, we can apply equiprobability arguments and calculate the probability of our having adopted the actual order which would coincide with the old lady's guesses, and so obtain a measure of the degree of coincidence we should have to invoke to explain away a sequence of correct guesses. Thus although we cannot at the outset assign any

† See above, Ch. VII, p. 118.

probability to the old lady's deeming a cup of tea pre- or post-lactationist, we can say that if we determine the order by drawing four cards out of a pack of eight, we can assign a 1 in 70 probability against our having happened to draw that order which would coincide with the old lady's actual guesses. In order to believe that the old lady lacks the power of discrimination, we have to believe that our experimental methods have been bedevilled by chance, and this is something which, beyond a certain threshold of improbability, we ought not to go on believing.

A similar methodological assumption is made in the so-called proportional syllogism, that if a proportion α of the members of a finite class have a characteristic, and if something is a member of the class but *nothing else relevant is known about it*, then the probability of its having the characteristic is α. For example, if $0\cdot2\%$ of Swedes are Roman Catholics, and Petersen is a Swede, then the probability of Petersen's being a Roman Catholic is $0\cdot002$. The proportional syllogism has confused philosophers because it can be interpreted sometimes as being synthetic, sometimes as being analytic, without the shift being recognised. We may take the italicised phrase, *nothing else relevant is known about it*, as giving us guidance on when to draw the conclusion; in which case an argument for equiprobability needs to be deployed, and the proportional syllogism becomes a special case of taking a sample, namely the case where the sample consists of only one member. Under pressure, however, we are liable to retreat into analyticity, and to understand the italicised phrase as stipulating that the required equiprobabilities obtain. The logic of the argument in either case is the same. Let us denote the property of being a particular named Swede by S_r; that is $S_r(s)$ is the propositional function that a Swede, s, is the rth one. Suppose there are altogether n Swedes, of which $n\alpha$ are Roman Catholics. Since the order of numbering is immaterial we can suppose that S_1, S_2, $\ldots$, $S_{n\alpha}$ describe the properties of being a particular Swede who is a Roman Catholic, and $S_{n\alpha+1}$, $S_{n\alpha+2}$, $\ldots$, S_n describe the properties of being a particular Swede who is not. Then for any Swede, s, $S_1(s) \vee S_2(s) \vee \cdots \vee S_n(s)$. Moreover, we are to assume that $\text{Prob}[S_r(s)] = \text{Prob}[S_t(s)]$, for $r, t = 1, 2, \ldots, n$, from which it follows that $\text{Prob}[S_r(s)] = 1/n$, and hence that $\text{Prob}[S_1(s) \vee S_2(s) \vee \cdots \vee S_{n\alpha}(s)] = n\alpha/n = \alpha$. If we do not secure this assumption of equiprobability by stipulating it, we justify it by reference to our method of encountering Swedes. Not only have we no reason to suppose that $\text{Prob}[S_1(s)]$ is any different from $\text{Prob}[S_2(s)]$ etc., but if it were, it would seem to show something strange about us. It is clear that much depends on the way I have come across Petersen. If I happen to be standing outside St. Ignatius' Church on a Sunday morning, I should not be surprised if my chance-met Swede turns out to be a Roman Catholic after all: whereas if I had a numbered list of all Swedes, and then drew a number out of a hat, it would indeed be surprising for such a selection-procedure to favour Roman Catholics, or to pick out any one Swede rather than any other.

A *sample* is a subclass available for investigation. We often cannot find out all we should like about every member of some class, either because we lack time or patience or because we cannot get hold of most members of the class to examine them. Instead, we examine only some members, and attempt to draw conclusions about the others on the basis of what we have found out about the some. We can find out what proportion of the sample are Roman Catholics, or what their average height is, or what the standard deviation of their height is: and from this we seek to infer what proportion of the whole population are Roman Catholics, what the mean or standard deviation of the height is. Two things can go wrong. The procedure for selecting the sample may be biassed: or the sample itself may just happen to be misleading.† However the sample is selected, it will be a proper subclass of the whole, and will therefore be characterized by some other feature in addition to that characteristic of the whole. Not every moth is examined but only those that are collected by a mercury-vapour lamp shining in a particular wood on a particular night. We are no longer dealing with all g's but only those that are also F, *i.e.* only with fg's. The probabilities we assign to propositional functions ranging in the universe of discourse of the g's will not in general be the same as we should assign to them ranging in the universe of discourse of the fg's. Only if $F(g)$ is independent of $H(g)$, can we say that $H(fg) = H(g)$. And of whether $F(g)$ is independent of $H(g)$ or not, the layman is often the proper judge. Statisticians are usually fairly fly in their sampling procedures and guard against

† I borrow the term from Ian Hacking, *Logic of Statistical Inference*, Cambridge, 1965, p. 128. See more fully, the whole of his Ch. VIII, esp. pp. 126–8.

all sorts of bias we had barely thought of: but the question of what could be a bias, of what factors could be relevant and could affect the answer obtained, is a question on which we are the authorities, not the statistician. The entomologist must say whether moths collected by a mercury-vapour lamp at night will be typical of moths in general or not. If I want to know the way that people are going to vote at the next election it is for me to say that picking names out of a telephone directory will produce a biassed sample, because only the rich have telephones: whereas it might be an acceptable way of doing research into colour blindness, because no correlation between that and wealth is at all likely. Even if the sampling procedure was quite respectable, the sample itself may be misleading. If I notice that all those interviewed in a university survey happen to be Old Wykehamists, I shall not infer from their all wanting to go into the Civil Service that everyone in the university does. However reputable the procedure, common sense may interpose a 'but' about the actual samples on which an inference is to be based. Nor need the statistician himself be deficient in common sense. They can avoid being landed by ill luck with certain types of misleading samples, by adopting a different standard of randomness. Instead of making a series of random selections, one man at a time, from the whole university, thus running a small but real risk of collecting an entire sampleful of Old Wykehamists, the statistician may choose a sample which shall contain the same proportion of Old Wykehamists, Old Rugbeians, Old Bristolians *etc.* as the university as a whole: then at least he will not be misled by that sort of coincidence. More formally, if in a universe of discourse g where we want to assign a value to $\text{Prob}[H(g)]$, and we suspect that K, L, M, are relevant to H, and we know the proportion of k's, l's, and m's among the g's, we arrange our sample $F(g)$ so that $\text{Prob}[K(fg)]$, $\text{Prob}[L(fg)]$ and $\text{Prob}[M(fg)]$ are the same as $\text{Prob}[K(g)]$, $\text{Prob}[L(g)]$ and $\text{Prob}[M(g)]$. But, of course, it may be a very disputable question whether some feature is or is not relevant, and which of those that are relevant should be mirrored in the sample. When we are dealing with human beings there is no limit to the number of factors that are relevant, and the only really representative sample is the whole population in its entirety. We may be content to argue from smaller samples, and often they will not in fact be misleading. But we can never be sure that they are not:

statistical procedures can save us from some errors, but not from all; and it is worth looking at actual samples to see if they happen to be in any serious way untypical.

Our last ploy is the destruction test. If the conclusion turned out after all to be false, which link in the argument would the statistician abandon first? If the argument is based on an assumption of equiprobability, it may be some principle of symmetry or indifference that would have to be abandoned: the sampling procedure was not as random as he thought. If the argument is Bayesian, we might be forced to posit some singular and hitherto unexpected hypothesis about the distribution of initial probabilities. If, as most often, there is a level-of-significance argument involved, we want to know what the level is and what sort of coincidence would have to have occurred if the conclusions in fact drawn were wrong. We need also, where level-of-significance arguments are involved, to ask which way the burden of proof lies. When the contraceptive pill first came in, a number of women died of thrombosis after taking it. Re-assuring statements were issued that the number of deaths from thrombosis among pill-takers was not "statistically significant". Many members of the public understood this to mean that only by a one-in-a-thousand chance could a harmful pill have produced so few deaths, rather than that the pill *might* be harmless, and that the deaths *could* be due to coincidence, and the chances against such a coincidence occurring were not as yet twenty-to-one. But it is a very different thing to be proved innocent beyond reasonable doubt and not yet to be proved guilty. Before passing a drug as safe, we are right to demand a 0·1% level of significance: but if I need already to invoke a one-in-nineteen coincidence to reconcile a belief that a drug is harmless with the observed facts, then I should make it clear, to myself and to everybody else, that this is how the case lies.

When we are dealing with levels of significance, it is easy to see what the alternative to accepting the statisticians' conclusion is—coincidence. We always can invoke coincidence, although, as we have seen, if we are to be reasonable, we cannot do so always. If we invoke coincidence, we may have to climb down when further evidence emerges; and until then will characteristically be unable to draw any conclusions at all from the evidence we originally had. Our position is not so much one of dissent from the statistician's conclusion as of agnosticism. With an equiprobability or a Bayesian argument we are more likely to have to put forward an alternative hypothesis, which in the event would be more unacceptable than the conclusion which the statistician offers. In either case the exercise is worthwhile. For by seeming to dissent, we shall discover the real strength of the statistician's position. It is worth putting up with some abuse in order to obtain the feel of his arguments. In the end, it is the non-statistician who has to decide. Statistical techniques may enable him to see which way he should decide in cases where the numerical evidence is confused and conflicting. But the branch of knowledge in which the decision is being taken is not statistics, but genetics, botany, ecology, medicine, sociology, economics, psychology, or psephology. Logic is useful to the historian, and enables him to think more accurately and write more clearly about history; but is no substitute for historical knowledge and understanding. And similarly statistical techniques, although like logic applicable in many disciplines, are themselves no substitute for the knowledge and understanding peculiar to each discipline. And therefore the non-statistician should never give his judgement entirely into the keeping of his statistical advisers. In ordinary life, where the non-statistician is not master of all non-statistical disciplines, it is a counsel of perfection, and often we are presented with statistical verdicts in fields in which we are not competent to hold alternative opinions. No rules of thumb can be adequate, but often we have to go by them: and I have found it wise to believe statistical results obtained in the physical, biological, and medical sciences, but not to rely on any statistical inferences in the social sciences, and particularly not where political issues are involved.

X

CONTINUITY AND DENSITY

PROBABILITIES are a continuous interpolation between the discrete truth-values True and False. The generalisation from discrete qualities to continuous quantities is characteristic of natural science. We start by classifying things by reference to qualities, features, characteristics which they either do or do not have. Tigers are carnivorous; horses are not. But soon we start making comparisons, and say that gold is heavier than lead; and go on to assign magnitudes, and say that the specific gravity of gold is 19·3. We do this not only with our descriptions, but with our explanations and predications. We start with simple causal statements—if a drop of sulphuric acid is put on a mixture of potassium chlorate and sugar an explosion will ensue: but go on to say *what proportion* of ingredients must go into the cake mixture, *how hot* the oven must be, and *how long* it will take to bake. More generally still, the paradigm laws of physics (although, importantly, not the only ones) are not so much *causal chains* as *causal cords*. They are not simply of the form

$$(x)[A_1(x) \;\&\; A_2(x) \;\&\; \cdots \;\&\; A_n(x) \mathbin{.} \supset Z(x)]$$

where we only have to decide *whether* a particular situation—x—is A_1 or not, A_2 or not, *etc.*: and *if* the answer is yes in each case, then we know that the feature Z will also be present.† Instead, they are of the form

$$z = f(a_1, a_2, \ldots, a_n; t)$$

where $a_1, a_2, \ldots, a_n, z$, are quantitative magnitudes, measured by real numbers, instead of qualitative features which can only be either present or absent, and f is a single-valued mathematical

---

† For a general account of simple—that is, discrete—causal laws, see J. R. Lucas, "Causation", in R. J. Butler, ed., *Analytical Philosophy*, Series I, Oxford, 1962, pp. 32–65.

function†, by means of which we can calculate the value of some variable z, after a period t has elapsed, given the initial conditions, $a_1, a_2, \ldots, a_n$.

We therefore want to make probability-judgements about quantities as well as qualities, and ultimately to generalise the concept of a causal cord to something more probabilistic.‡ The first stage is to consider the quantitative analogue of a propositional function. A propositional function—e.g. $F(g)$—has two terms: one—g—indicating the subject or universe of discourse, the other—F—saying what feature is or is not being ascribed. It is only the latter that admits of quantitative generalisation. We can say of this oven how hot it is or how heavy it is, but it remains this oven, neither more nor less, that we are speaking of. Subjects are necessarily discrete and do not admit of comparatives, or *a fortiori* of quantities.§ Although we can use numbers to refer to subjects, as we do when we refer to places on a map by means of a grid reference, each place is referred to as an individual topic of discourse about which something is being said. And it is only the something that is being said about it which can be a magnitude. We can talk of an oven being black, of its having a 90% probability of being black, of its being at 400 °F, or of its having a 90% probability of its being at 400 °F: but we cannot be talking 0·1 about an oven and 0·9 about something else. We must talk about what we are talking about, and nothing else; otherwise communication breaks down. We can hedge on our answers, but not on our subjects of discourse. Our conversation would be hopelessly at cross-purposes if our references were so allusive we had to be understood as always talking about several things at once, none

---

† Normally we write $x_1, x_2, \ldots, x_n$ instead of $a_1, a_2, \ldots, a_n$: and z will be the value of one of these, say a_1, after a period of t has elapsed, and is written x_1'. So altogether there is a set of n functions:

$$x_1' = f_1(x_1, x_2, \ldots, x_n; t)$$
$$x_2' = f_2(x_1, x_2, \ldots, x_n; t)$$
$$\cdot \quad \cdot \quad \cdot \quad \cdot \quad \cdot \quad \cdot \quad \cdot$$
$$x_n' = f_n(x_1, x_2, \ldots, x_n; t).$$

‡ In Ch. XII, pp. 193–5.

§ Compare Aristotle, *Categories*, Ch. V, 3b12–13, 33–4. ἄτομον γὰρ καὶ ἐν ἀριθμῷ τὸ δηλούμενόν ἐστιν ... δοκεῖ δὲ ἡ οὐσία οὐκ ἐπιδέχεσθαι τὸ μᾶλλον καὶ τὸ ἧττον. The thing shown (*i.e.* referred to be a subject term) is individuated as a discrete entity ... it seems that a substance does not admit of degrees.

of them whole-heartedly. Hence it is only the predicate-variable F of the propositional function $F(g)$ that we generalise, not the individual-variable g. We shall talk about the probability of an oven's being at 400 °F or the probability of an adult Englishman's being 6 ft. tall, and thus replace F by some continuously varying magnitude: but we shall leave g alone.

As soon as we introduce quantities rather than qualities, we run into problems of exactitude. 'Do you mean *exactly* 400 °F?' we ask. 'Do you mean exactly 6 ft. tall, or only 6 ft. tall or over?' If the former, then, a few special cases apart, the probability is going to be vanishingly small. By measuring exactly enough, almost all those who do not fail through being less than 6 ft., will be found to be more than 6 ft., and therefore not exactly 6 ft. It is a standard difficulty in mathematics, and has a standard solution: instead of talking of *probabilities*, we talk of *probability-densities* and integrate. Essentially we do not attempt to talk of the probability of being *exactly* 6 ft., but only of being *about* 6 ft., and would like to assign a probability value not to a man's height being *at* 6 ft. but to its being at 6 ft. *or thereabouts*. The obvious difficulty is that we have not specified how broadly we are going to take "about" or "or thereabouts": and clearly the probability depends on that—it would be much higher if we took it as being within 4 inches, than if we were taking it to within $\frac{1}{2}$ inch. Roughly speaking, the probability will be proportional to how broadly we are taking 'about'. And so for mathematical purposes we divide the probability by the 'about', and call the result the *probability-density*. Then, as long as the probability-density is not changing very much for different heights (which it will not, if the difference is very, very small), the probability of a man's height being *between* 6 ft. and $6 + \delta x$ ft. inclusive, where δx is very, very small, will be the product of the probability-*density at* 6 ft. and δx. So, if we replace the F of $F(g)$ by D—for density—and specify the height we are interested in—6 ft.—we shall have $D(6)(g)$—as the probability-*density* of a g's being 6 ft., and then the *probability* of a g's being between 6 ft. and $6 + \delta x$ ft. inclusive will be $D(6)(g) \times \delta x$; and for δx which are not very, very small we can use integrals, and say that, e.g., the probability of a g's being between 5 ft. and 7 ft. will be $\int_5^7 D(x)(g) \, \delta x$. $D(x)(g)$ is a cumbersome formula, and normally mathematicians leave out the g—and, indeed, forget about it, since it is not mathematically interesting, not being able to vary continuously. We shall follow them for the present, but later† the universe of discourse will be important to our argument.

The generalisation from probabilities to probability-densities is natural and attractive: but it is not inevitable. We can have probabilities, and not probability-densities, assigned to individual points. But if we do, we can assign finite (non-infinitesimal‡) probability-values to only a finite (non-infinite) number of such points; for else the total probability of the logical sum of all the alternatives would be greater than unity. If we do not want to have to assign the value zero to all but a finite number of values of a continuous variable, we must deal in probability-densities rather than probabilities. For we can assign finite probability-*densities* to an infinite number of values without infringing the normalisation rule that the sum of all the probabilities must not exceed unity, since, with probability-densities, we have to multiply not by the *number* of values (which may well be infinite) but by the "volume" (which may well be small). Thus we can assign probability-densities to infinite numbers of points, and so to all the points in a continuous interval. Therefore if we want to have any sort of probabilistic assignment which varies continuously, it must be an assignment of probability-densities, which can then be integrated over any given range or "volume" to determine the probability of having values lying within that range or volume. The normalisation requirement, which for probabilities was that the probability of the logical sum of all possibilities would be unity, becomes for probability-densities that the integral of the probability-density over the whole space shall be unity. This is the natural extension to the continuous case. But provided we are prepared to accept discontinuity, we *can* assign probabilities, and not probability-densities, to individual points; only, we can assign non-infinitesimal values to only a finite number of points, and must assign the value zero to all the rest.

Sometimes we may want to talk both of probabilities and of probability-densities in the same breath: and this leads to difficulties. It is not absurd to say that the probability of a certain magnitude, x say, taking the value 2 is 0·5, the probability of its

† In Ch. XII.
‡ Members of a set are non-infinitesimal if they are all greater than some one member, itself greater than zero.

taking the value 3 is 0·25, and the probability of its being negative is 0·25, with a probability-density given, for negative values of x, by the function $(4 - x)^{-2}$. This in itself is reasonable and unexceptionable. The difficulty comes in trying to talk of probabilities at a point in terms of probability-densities. For the range or volume occupied by a point, say $x = 2$, is zero; and therefore the probability-density must be infinite, if we are not to have to assign a zero probability to it too. But even with an infinite density we cannot be sure we are safe, for infinities are treacherous concepts. We can, as Dirac did, create a function by *fiat*, with the three desired properties: that it is infinite at one point; that it is zero at every other; *and* that it has a definite, finite non-zero, integral (usually, in fact, set equal to unity). But such a function is disreputable in the eyes of the mathematicians, for no function has quite been produced that actually has (rather than tends towards having) quite these properties. What is required is not a mathematical function but a conceptual adjustment to the σύζευξις of the two modes of discourse, that of probabilities and that of probability-densities. We want to be able to assign probabilities occasionally to points as well as usually to ranges or volumes. In the latter case we assign probability-densities to points, and then integrate over the relevant range or volume. In the former case there is no need to do this; but if, for the sake of uniformity, we want to talk all the time in terms of probability-densities, then we will need to use a Dirac delta function, as a *façon de parler*, to accommodate the occasions when we are really assigning to a particular point a probability and not merely a probability-density.

Except in quantum mechanics, we never need deal with Dirac delta functions. Invariably when we deal with continuous magnitudes (which we usually represent by the X-axis, or abscissa, on a graph), we assign to them only probability-densities (which we usually represent by the Y-axis, or ordinate, on a graph); probabilities themselves are measured by areas—the area under the curve between two ordinates represents the probability that the value lies between the two corresponding points on the X-axis. There are many probability-density distributions with useful applications, but the only one of philosophical importance is the Normal—or Gaussian—Distribution, which we often come across in naturally occurring populations, and for which some *a priori* justification can be given. The Normal Distribution is usually expressed by the formula

$$\frac{1}{\sigma\sqrt{2\pi}}\exp\left[\frac{-(x-\mu)^2}{2\sigma^2}\right], \quad \text{or} \quad \frac{1}{\sigma\sqrt{2\pi}}\,e^{\frac{-(x-\mu)^2}{2\sigma^2}}$$

as it is sometimes written, where μ is the mean, or average, value of the quantity, and σ is the standard deviation.†

Non-mathematicians may take fright at the formula

$$e^{\frac{-(x-\mu)^2}{2\sigma^2}},$$

and wonder how one can multiply the mysterious number e, which cannot be expressed as any fraction, by itself a negative, and possibly irrational, number of times. It is an unnecessary worry. The exponential function is defined not in terms of e, but as the inverse of log, which is itself defined as that function of, say, x which will have $1/x$ as its differential coefficient. Essentially, a simple exponential function is one in which the gradient (measured by the differential coefficient) is equal to the value of the function. In our case the standard rules for differentiating a function of a function tell us that the gradient will be the product of the function itself with the differential of the exponent: that is,

$$\frac{d}{dx}\left[\exp\left[-\frac{(x-\mu)^2}{2\sigma^2}\right]\right] = \left[\exp\left[-\frac{(x-\mu)^2}{2\sigma^2}\right]\right] \times \left[\frac{(\mu-x)}{\sigma^2}\right];$$

that is to say, the ratio of the gradient to the value of the function itself is proportional to $\mu - x$, the distance from the mean μ. We also notice that since $(x-\mu)^2 = [-(x-\mu)]^2$, the graph will be symmetrical about the mean. In fact, the graph looks like this:

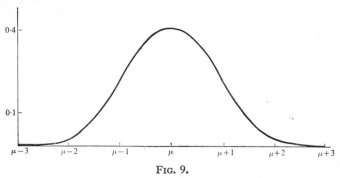

FIG. 9.

† See above, Ch. IX, pp. 166–7.

The reader should note that, as in most textbooks, the scale along the vertical axis has been magnified four or five times for the sake of clarity. It is reminiscent of the graph of the Binomial Distribution when we have taken care to adjust the scale suitably,† and is indeed the continuous analogue of it. As with Bernoulli's Theorem, a direct but somewhat unintuitive proof can be given with the aid of Stirling's formula. I shall give a more intuitive proof, slightly glossing over difficulties about the scale as the n of the Binomial Distribution tends to infinity and we slide from the discrete to the continuous case. Essentially, we find a rather simple formula for the gradient, which we can integrate quite easily to give the required result. As we have seen,‡ the ratio

$$\frac{{}^nT_r}{{}^nT_{r-1}} = \frac{(n - r + 1)\,\alpha}{r\,(1 - \alpha)}$$

and therefore ${}^nT_r$ reaches its maximum value when $r = l$, where $(n + 1)\alpha - 1 \leqslant l \leqslant (n + 1)\alpha$. We now seek to calculate from this ratio an expression for the gradient in terms of the "distance" from the middle. Writing $l + u$ instead of r, we have the gradient from ${}^nT_{l+u-1}$ to ${}^nT_{l+u}$ is proportional to $[{}^nT_{l+u} - {}^nT_{l+u-1}]$. Instead of calculating the difference directly, we calculate it in terms of the ratio, which we have already. We calculate

$$D = \frac{[{}^nT_{l+u} - {}^nT_{l+u-1}]}{{}^nT_{l+u-1}} = \frac{{}^nT_{l+u}}{{}^nT_{l+u-1}} - 1 = \frac{(n - l - u + 1)\alpha}{(l + u)(1 - \alpha)} - 1$$

$$= \frac{(n + 1)\alpha - (l + u)\alpha - (l + u) + (l + u)\alpha}{(l + u)(1 - \alpha)}$$

$$= \frac{(n + 1)\alpha - l - u}{(l + u)(1 - \alpha)}$$

now $l \doteqdot (n + 1)\alpha$, hence the numerator $\doteqdot -u$, and the denominator $\doteqdot ((n + 1)\alpha + u)(1 - \alpha) \doteqdot (n\alpha + u)(1 - \alpha)$.

Thus far the argument is perfectly rigorous: when we slide over into the continuous case, we find it difficult to see what is the right scale for measuring this distance. If we just assume that each Binomial term is a unit distance from the next, then, as we began

† Compare Ch. V, figures 3 and 4, pp. 77 and 78.
‡ Ch. V, p. 79.

to see in Chapter V, the figure will flatten out as n increases.[†] If, on the other hand, we measure distance by proportion, *e.g.* δ, we have the peak becoming more and more of a spike as n increases— this is what lies behind Bernoulli's Theorem.[‡] What we want now is to choose our coordinates so that the position of the peak does not alter with increasing n, nor the measure of distance away from the peak. We have secured the first by writing our expression in the form $l + u$, and then eliminating l; to secure the second, we stipulate that the standard deviation— σ —shall not depend on n. The standard deviation constitutes a reasonably relevant measure of the scale along the abscissae, the X-axis, of a probability distribution. For a Binomial Distribution, the standard deviation can be shown to be $\sqrt{n\alpha(1 - \alpha)}$.[§] It is because it varies not with n but only with $\sqrt{n}$ that figure 3 in Chapter V flattens out for increasing n; and it is because it varies with $\sqrt{n}$, and is not a constant independent of n that figure 4 in Chapter V becomes steeper and more concentrated round the middle for increasing n. We therefore require our measurements along the abscissae neither to vary directly with n nor to be wholly independent of it, but to vary with $\sqrt{n}$. Hence as n increases, u increases only as $\sqrt{n}$, not as n, does. Hence in the denominator $((n + 1)\alpha + u)(1 - \alpha)$, the u is smothered by the $(n + 1)\alpha$ as $n \to \infty$. The whole fraction approaches

$$\frac{-u}{(n + 1)\alpha(1 - \alpha)} \doteqdot \frac{-u}{n\alpha(1 - \alpha)} = \frac{-u}{\left(\sqrt{n\alpha(1 - \alpha)}\right)^2}$$

but $\sqrt{n\alpha(1 - \alpha)} = \sigma$, hence the fraction becomes $-u/\sigma^2$. It is now entirely straightforward to integrate. The fraction D was proportional to the gradient divided by the value of the fraction itself at that point: *i.e.*

$$\frac{f'(u)}{f(u)} = \frac{-u}{\sigma^2}.$$

Hence

$$\log f(u) = \frac{-u^2}{2\sigma^2} + C$$

[†] Ch. V, figure 3, p. 77.   [‡] Ch. V, figure 5, p. 80, and figure 4, p. 78.
[§] See below, Appendix 1, pp. 211–12 for a proof of essentially this result, but in terms of frequencies rather than absolute magnitudes.

where C is an arbitrary constant

$$\therefore f(u) = A \exp \left[\frac{-u^2}{2\sigma^2}\right] \quad \text{where } A = \exp [C].$$

We calculate A by integrating again and normalising. A very neat way of performing the second integration is to take the square of $f(x)$, and shift from Cartesian to polar coordinates.†

$$\left[\int_{-\infty}^{\infty} f(u) \, du\right]^2 = A^2 \int_{-\infty}^{\infty} \int_{-\infty}^{\infty} e^{\frac{-v^2}{2\sigma^2}} \times e^{\frac{-w^2}{2\sigma^2}} \, dv \, dw$$

$$= A^2 \int_{-\infty}^{\infty} \int_{-\infty}^{\infty} e^{\frac{-(v^2+w^2)}{2\sigma^2}} \, dv \, dw$$

$$= A^2 \int_0^{\infty} \int_0^{2\pi} e^{\frac{-r^2}{2\sigma^2}} r \, dr \, d\theta$$

$$= A^2 \int_0^{2\pi} \left[e^{\frac{-r^2}{2\sigma^2}} \times -\sigma^2\right]_0^{\infty} \, d\theta$$

$$= A^2 \int_0^{2\pi} \sigma^2 \, d\theta = A^2 2\pi\sigma^2.$$

$$\therefore \int_{-\infty}^{\infty} f(u) \, du = A\sigma\sqrt{2\pi}.$$

But this must be equal to 1, as it represents the sum total of the probabilities of all the possibilities. So $A = 1/(\sigma\sqrt{2\pi})$. In our derivation we have measured u from the summit; for the more general case, where the summit is at $x = \mu$, we replace u by $x - \mu$, and then we have $f(x)$ in the form required.

The Normal Distribution is much loved by statisticians on account of its mathematical properties. It is interesting philosophically because we can see why it might well be the distribution of the probability-densities of some naturally occurring magnitude. Consider height. How tall a man is depends on *many* factors: how tall his parents are, how well he was fed in youth, how much exercise he had, how many illnesses he suffered from, *etc.*, *etc.*, *etc.* We do not know what all the factors are, nor how many they are. But we believe that they are many, and that barring a few catastrophes, like infantile paralysis, no single factor has a decisive or very large effect. It is reasonable to assume, then, that a man's height is more or less determined by a large number of factors each having by itself a comparable effect. If further we make the assumption—not always justified—that the factors are

† See W. Feller, *An Introduction to Probability Theory and its Applications*, 2nd ed., Vol. I, New York, 1957, pp. 164–6. The rigorous justification of the first line needs fairly sophisticated argument.

all independent of one another—that is, if both A and B are present, the effect will be the sum of the effect due to A and the effect due to B—and that these factors are themselves randomly distributed, we shall have an analogue to our Bernoulli trials with coins. Each man is like a large number of tosses of a coin. For with each of the relevant factors, he either does or does not possess it, and therefore either is or is not that small amount taller than a man without that factor. Assuming, for simplicity's sake that the probability is the same for each factor, it is clear that we shall have a Binomial expression for the probabilities of the presence or absence of factors in a man, and that the proportion of men, as of sequences of coin-tossing, having different numbers of favourable factors present will follow the Binomial Distribution. Likewise their heights. And where the population is large, and since height is anyhow a continuous magnitude, we may properly approximate to a Normal Distribution instead.

These assumptions can be weakened without altering the result. Clearly, we do not need to know that both the probability of each factor and the increment of height resulting from it are equal, but only that their product is, provided the effect of the factor is not disproportionately large, compared with the effects of many other factors taken together. Nor need all the expectations of increment be even approximately equal, so long as the larger ones are approximately equal and fairly numerous. Nor need all the factors be independent: if they interact strongly, we can consider not the presence or absence of each factor separately, but of the relevant combinations of them. Provided there are still a large number of such combinations that are independently relevant to how tall a man is, the argument will still hold. In brief, we should expect to get something fairly like a Normal Distribution for the values of a single continuous variable—such as height—in a large population, especially in biology whenever we are measuring some quantitative attribute of organisms. For organisms are homeostatic systems. They survive because changes in their environment do not produce correspondingly large changes in them—the salinity of our body fluids varies less than the salinity of our food and drink. But they cannot be completely independent of their environment: on the contrary, in order to remain relatively constant in some respects, they must make many adjustments in others. Every organism therefore will be susceptible to a large number of environmental influences, but the effect of each will with some exceptions—*e.g.* stimuli on the sense-organs of animals—be fairly small. Moreover the range of these influences is so large that it is reasonable to assume that they occur independently of one another. Thus, so far as environment is concerned, we should expect a Normal Distribution.

Inside the organism our arguments must be less *a priori.* It is only a contingent truth that many organisms are very complicated and so involve many factors: it is even more contingent that there is a large measure of decentralisation in complicated organisms, and that their physiology consists of a large number of relatively stable and independent biochemical reactions. Nevertheless it is fair to adduce these general biological considerations to support the assumption of there being a Normal Distribution, when we are trying to discover the height of Englishmen or the weight of field-mice in Ireland. More telling, however, is genetic theory. Once we have sexual reproduction and the transmission of inherited characteristics, each individual's genetic inheritance will be analogous to a Bernoulli trial of throws of pairs (because there are two gametes) of (possibly two- or possibly many-faced) dice. And this gives strong theoretical support to the assumption of Normality. Moreover, in practical contexts, it is very likely that we shall have noticed the special cases. The natural historian will have noticed a difference of sub-species long before the biometrician has discerned a difference of vital statistics. Common sense prevents us putting unfairly abnormal questions to the statistician. And so, by the same token, the statistician is entitled to make the reasonable, but not watertight, assumption of there being a Normal Distribution of the quantity in question, *unless . . .* unless certain rather special conditions obtain.

Unless . . . We shall not get a Normal Distribution if there is one factor which has far greater effect than all the others. Mendel's own experiments with tall and dwarf peas provide an example; one genetic factor makes a very large difference to the height of peas, and we do not get anything like a Normal Distribution of heights, but at best the superimposition of two Normal Distributions around two 'peaks'. Again, we shall not get a Normal Distribution if the independence condition is too much eroded: if many of the factors are genetically linked, or if the population is not genetically well-mixed. Many genes are sex-linked, and the effect of many others is influenced by sex, and we are not surprised that the distribution of height is different for men and for women however much they inter-marry. Nor are we surprised that the distribution of skin-pigmentation—if that can be assessed on a quantitative scale—is not Normal in South Africa. Very often—not only in human populations—there are subtle barriers against complete genetic intermixture; it is the way that species begin to differentiate themselves:† and in such cases an assumption of Normality may be wrong. Nevertheless, these are special cases. If we assume that the distribution of some continuously varying feature in a large population is Normal, we may always discover that we were wrong, and have to replace our assumption by something better. But in the absence of any reason to the contrary, it is itself a reasonable assumption.

One further philosophical objection apart,‡ these arguments are adequate. The onus of proof is on the man who does not expect there to be a Normal Distribution, rather than on him who does. And if we can assume a Normal Distribution, our samples become very much more informative than in the discrete case. For the sample will show not only what its mean, or average, value is, but how widely dispersed the different particular values are around this mean. And from this we can infer both the limits within which the mean of the whole population must lie, unless our sample was, by an implausible coincidence, remarkably untypical, and what sort of variation we may expect. We are much better placed when we have a sample in which we are measuring quantities than one in which we are only counting the occurrence of qualities: for in the latter the only information we obtain is the frequency of occurrence of some quality, whereas in the former we learn also the dispersion of values in the sample. Once again, it is helpful to view *each* member of a population as analogous to a *whole* Bernoulli trial, a whole series of tosses of a coin. Measuring the height of a large number of individual members of a population is like being given the frequency of heads in a large number of Bernoulli trials: which, of course, would give us much more information about the basic probability than only one such trial. Statisticians are thus able to make much more precise

† G. S. Carter, *A Hundred Years of Evolution*, London, 1957, pp. 152–6.
‡ See below Ch. XI, p. 187.

estimates than the ones based on a simple Bernoulli trial, discussed in Chapter V. But the logic remains the same, and depends essentially on its having to be an implausibly improbable coincidence that a Normal Distribution of a certain posited mean and certain posited standard deviation should yield a sample with this observed mean and this observed standard deviation.

XI

BIOLOGICAL PROBABILITIES AND IMPERFECT INFORMATION

ONE philosophical difficulty remains. Biology is felt to be a determinist science, in which complete explanations in purely physical terms are in principle possible, even if in practice seldom obtained. Therefore, it cannot correspond to the truth to ascribe a probability to the possession of a certain characteristic by a particular member of some population. It may be a useful *façon de parler*, but in reality every member either does or does not have the characteristic in question, and to talk in terms of probability is at best a *pis aller*. It is not a problem peculiar to biology, although most acute there. It occurs also in thermodynamics, and troubles the ordinary man when he ponders the mystery of the luck of the draw in cards and the chances apparently yielded by dice. In each case our use of probabilistic discourse seems to be due to our lack of perfect information: some of the parameters are hid from our eyes, and so we fall back on judgements of probability because the certainty which is there happens to elude us. God, to whom all things are known, could never make a probability judgement: Einstein was right to refuse to believe in a dice-playing God, for, on this view, it would be a contradiction in terms.

But probability, like truth, does not attach to things, only to propositions and propositional functions. And while the thing is independent of what we say or what we think about it, the proposition or propositional function is not.† Under one description a thing may be said to have a probability of possessing some characteristic, whereas under another it would be plainly false to say that it had. If we specify (supposing, for the sake of argument, that we could) an individual member of a given population completely, then we can only affirm true or false propositions about it, Leibniz-wise: we can only say that it possesses the characteristics

---

† See above, Ch. VI, pp. 106–8.

it does possess, and does not possess those it does not—and these will all be true propositions—or that it possesses the characteristics it does not possess and does not possess those it does—and these will all be false propositions. But if we specify the subject less fully, then we may have occasion to make judgements of probability. To take Jeffreys' example,† if we describe Smith merely as an Englishman, we shall assign one probability-value to the propositional function that such a man has a blue right eye: if we describe Smith as an Englishman with a brown left eye, we shall assign another probability-value to the propositional function that such a man has a blue right eye; whereas if we give a fully detailed description of Smith, even down to the colour of both his eyes, then it must be either that he does have a blue right eye or that he does not, and in the former case the probability is 1 and in the latter it is 0, and no value in between could possibly be given. Similarly, provided we describe an individual member of a population simply as that, we can assign a probability-value other than 1 or 0 to the propositional function that in such an individual a given factor should be present. It is a question of description rather than of knowledge. If we apply a description which is incompletely specified, and therefore could be applied to other individuals as well, then in talking of the probability of *such* an individual having some characteristic we are talking in the universe of discourse of all such individuals: and in that universe the propositional function may well have a probability-value other than 1 or 0. It is not because we do not know whether an individual has a certain factor present or not, but because we want to consider him not just as an individual, but *qua* member of a population, that we use probabilistic rather than black-and-white, Yes-or-No, true-or-false language.

Are we entitled to describe incompletely? Insurance agents would say we are not. The doctrine of *uberrima fides* requires frank and full disclosure of all facts which might be material: an insurer would never insure a man under one description if, under another available description, a firm prediction could be made. Surely the biologist is not entitled to throw away information any more than God could forget his omniscience. But the biologist is not betting; he is generalising. He is not setting exaggerated store by the outcome of a single event; and therefore has no motive for

† See above, Ch. IV, p. 50.

minimising losses by examining that single event for every relevant detail. It is not the particularity but the typicalness of the
sample that he values. He wants to know about the population as
a whole, and has selected or is considering the individual only *as* a
member of that population; and therefore needs to use a general,
and hence incomplete, description which will cover any other
member of the population equally well. His purposes are different
from an insurer's. The insurer wants to have as much information
as possible in order to minimise losses on each particular contract:
the biologist wants to drop irrelevant particulars in order to maximise knowledge of the general.

Nor is it only a question of convenience in view of different
purposes. Rather, it is built into the very notion of a biological
science that it has a special criterion of relevance. If we were to
give a complete state-description in physical terms together with
the equations for working out its subsequent development, we
should not have given a fuller or better account of a biological
organism, but a worse. It would be useless. We should not be
able to see the wood for the trees. The life history of each oxygen
atom would be traced out in full detail, but there would be nothing to distinguish its course through the environment outside the
organism from its reactions in the physiological processes within
the organism. And that is the distinction the biologist wants most
to draw. He does not care what happens to each particular oxygen
atom in the environment. So long as there is oxygen available, one
atom is as good as another. Similarly inside the organism. It is
only occasionally that he wants to trace the course of particular
atoms, and then only against a backgound of a biological structure not defined in purely chemical terms. Normally he will
identify the leaves and petals and stamens, or the nervous system
and kidneys, without reference to any particular atoms; and only
then will he, perhaps, use a few radioactive isotopes to follow in
detail their biochemical budgets. But the first, second and third
task of the biologist is to pick out features which are biologically
relevant, and to discount the rest. Organism and organ, respira-
tion and excretion, reproduction and death, species, sub-species
and deme—these are the biologist's categories. If he was given
some fuller description, he would have to labour to re-introduce
these categories rather than the others he had been given. If, so
to speak, he was given on a graph the world-lines of every atom,

he would be inking in round those convolutions and knots that were living organisms, and rubbing out the lines that obscured his picture. Even if he knew them, they would be irrelevant to his concern. He need not be ignorant of them, but must ignore them if he is to get through to the more general features which are his interest.

It is not peculiar to biology. We offered a similar justification for using dice or cards in games of chance.† In arguing from the observed results of Bernoulli trials to the probability of the propositional function involved, we needed to neglect the order and concentrate only on the relative frequency.‡ The probability of a particular result under a description which specifies the order is different from the probability under a description which does not: and yet the latter is more informative than the former, for it gives the answer to a more general and more interesting question. So too with populations. They are a legitimate subject of research: hence an individual variable ranging over the members of a population is also a legitimate subject term for a propositional function, to which a probability may properly be assigned. We are entitled to talk about whole populations all in one breath: there are many features, genetic and otherwise, which it would be untrue either to say that the population did, or to say that it did not, possess them. In such cases probabilities fill the gap, and it is entirely legitimate to use them.

A similar argument applies in thermodynamics. It is not, as many expositors make out, that we cannot plot the positions and velocities of individual molecules, but that we do not want to. The impact of an individual molecule does not signify: only if there are lots and lots do they constitute a pressure. And since pressure is what we are interested in, we need to use statistical mechanics in order to talk about lots and lots and lots compendiously, briefly *and relevantly*.

The tenor of these arguments is to give a logical account of probability in terms of incomplete specification instead of an epistemological account in terms of imperfect information. 'Information' is itself an unfortunate word. It sounds as though it were a good thing which a man could not have too much of. But that is to commit the Dover fallacy§: when it relates to the speci-

† See above, Ch. VII, pp. 121–2.
‡ See above, Ch. VIII, pp. 131–3.   § See above, Ch. VIII, p. 128.

fication of the universe of discourse, the more information we use the more narrowly are we restricting the universe of discourse; so that an increase of information means that the question is narrower rather than that the answer is larger. What the right questions to ask are depends on the context, and a considerable amount of argument may be needed to show that some universe of discourse is a useful, or even intelligible, one. But the argument does not have to be that in view of our ignorance we can no other. Probabilities are not a second best, to be used when knowledge fails, but a good thing in their own right, which sometimes may be preferable to a fuller knowledge that is actually available.

Nevertheless a certain unease remains. Although probabilities do not have to give way to knowledge, and may for some purposes be preferred, yet they would seem to be still somehow derivative and secondary. In thermodynamics we may not need all the information which a complete state-description would give, but if we had it—and enough patience—we could derive the thermodynamical description: and perhaps a biologist could extract his descriptions too. Completeness of specification has a natural edge over incompleteness: irrelevance is popularly esteemed a much lesser vice than reticence. The ultimate description of the world, we feel, must be a complete one, and probabilistic discourse must be a mark, if not of our limited knowledge, then of our limited purposes.

XII

OBJECTIVE PROBABILITY AND PHYSICS

THE Subjective Theory of Probability, although open to the many criticisms we have made in Chapters II and IV, is more attractive when metaphysical issues are invoked. A philosopher might concede our point about the meaning of probability statements and allow that the *concept* of probability was objective, while yet insisting that there were no genuinely objective probabilities, and that the concept was inapplicable in any metaphysically fundamental level of discourse. In metaphysics an assumption of perfect information is easily made; often, as we saw, on the theological grounds, that God, to whom all things are known, could never make a probability-judgement. Even with the modification of the last chapter, probabilities are in a sense subjective in that they depend on our choice of universe of discourse, and that if we chose to say as much as we could about each individual thing separately then, ideally, we should need no judgements of probability. Even if God could have occasion to make probability-judgements, He would never be obliged to. And therefore Einstein was still right to believe that a dice-playing God, would, on most traditional views of omniscience, be a contradiction in terms. *Per contra*, if we adopt a fundamentally objective view of probability, it will have considerable metaphysical consequences.

The metaphysics of perfect information is best exemplified in Newton's philosophy of nature, which also takes a high view of substance and a somewhat low view of time. These views, I shall argue, are inter-related; and if we adopt a fundamentally probabilistic philosophy of nature, as the success of quantum mechanics suggests we should, we shall no longer be able to think of the fundamental stuff of the world as being a number of numerically distinct atoms, corpuscles, or point-particles, and we shall not have to think of time as being homogeneous and without any natural direction. These results depend on the fact that although we can ascribe probabilities to singular propositions as such, the full theory requires us to ascribe probabilities to propositional functions or propositions regarded merely as instantiations of propositional functions, and the two terms, F and g, in a propositional function, $F(g)$, play essentially different, and non-interchangeable, *rôles*.† It is because we cannot generalise to a continuous analogue of g in the way we can for F that our notion of primary substance breaks down: and it is because we cannot negate g, and so cannot contrapose $F(g)$ in the way we can $F(x) \supset G(x)$, that the rule of time-reversibility does not apply.

Quantum mechanics is probabilistic through and through. Not only is it, but Pauli‡ and perhaps von Neumann§ have given us good reasons for supposing that it must be. It is essentially unlike thermodynamics. Thermodynamics, although itself probabilistic, is compatible with, and indeed derivable from, a fuller, and completely determinist, system. One can always envisage an argument in statistical mechanics being filled out to give an exact mechanical account of every molecule involved. It may be tedious, it may be irrelevant, but it is not inconsistent to do it. But with quantum mechanics it would be inconsistent to suppose that there could be "hidden parameters" which would fill it out to a completely determinist system. Of course it remains formally on the cards that quantum mechanics is an inadequate theory which will be superseded by some other, determinist theory: but at the very least, we must be prepared for the possibility that the fundamental theories of physics are essentially probabilistic.

It would be entirely in line with the generalisation from discrete to continuous variables we have already noted. The paradigm

† See above, Ch. IV, esp. pp. 47–52, 56–65, Ch. VI, esp. pp. 104–8, Ch. VIII, pp. 126–7, Ch. X, pp. 175–6.

‡ For a very brief account, see Allan M. Munn, *Free-Will and Determinism*, London, 1960, Ch. VI, p. 161; or J. R. Lucas, *The Freedom of the Will*, Oxford, 1970, §20.

§ John von Neumann, *Mathematical Foundations of Quantum Mechanics*, tr. Robert T. Beyer, Princeton, 1955, Ch. IV, §2, esp. pp. 323–8. Allan M. Munn, *Free-Will and Determinism*, London, 1960, Ch. VI, pp. 159–61. But see also D. Bohm, "A Suggested Interpretation of the Quantum Theory in terms of 'Hidden' Variables", *Physical Review*, LXXXV, 1952, p. 166; *Causality and Chance in Modern Physics*, London, 1957. For a criticism, see N. R. Hanson, "On the Symmetry Between Explanation and Prediction", *Philosophical Review*, LXVIII, 1959, pp. 353–6.

of a natural law in classical physics is a functional dependence, in which we have n functions connecting n variables, together with a parameter representing time:

$$x_1' = f_1(x_1, x_2, \ldots, x_n; t)$$
$$x_2' = f_2(x_1, x_2, \ldots, x_n; t)$$
$$\cdot \quad \cdot \quad \cdot \quad \cdot \quad \cdot \quad \cdot \quad \cdot$$
$$x_n' = f_n(x_1, x_2, \ldots, x_n; t)$$

In the $(n + 1)$-dimensional phase-space of the n variables together with time, these equations will, for any set of initial conditions $(x_1^0, x_2^0, \ldots, x_n^0)$ at a given time t_0, define a continuous curve, which represents the development through time of a system with those initial conditions. This we call a causal cord, as it is a generalisation of a causal chain.† But although the discrete qualitative factors $A_1, A_2, \ldots, A_n$, which could only be either present or absent, have been generalised to quantitative magnitudes $x_1, x_2, \ldots, x_n$, which can vary continuously, the functional dependence remains discrete in one respect still: given a set of initial conditions and a definite value of the temporal parameter, any set of values for the variables $x_1, x_2, \ldots, x_n$ *either* does *or* does not satisfy the equations that define the functional dependence. That is to say, in the $(n + 1)$-dimensional phase-space, any given point *either* does *or* does not lie on the curve representing the development of a particular system. We are still dealing with world-lines, which either do or do not pass through a given point. It is therefore a natural extension of our programme of generalising from the discrete to the continuous case, to attempt, so far as we can, to generalise from points and lines to densities and their linear analogues. Just as we expect causal chains to be generalised to causal cords, so we should expect the cords to fuzz out to something more like vapour trails. But in so doing, they will lose some topological features that have played a crucial part in our metaphysics of science hitherto. We shall lose both our criterion of identity and our two temporal principles of date-indifference and direction-indifference. The criterion of identity for material substance, idealised as Newtonian atoms, is spatio-temporal continuity. But continuity disappears when we replace determinate values by a probability-density distribution. Given a particle at one time, a particle at a later time is identical with it if

† See above, Ch. X, pp. 174–5.

and only if there is a continuous path representing the particle's position which leads from the one to the other. It is a definite question, with a Yes-or-No answer. But there is characteristically no Yes-or-No answer as to whether a particular position is "on" a probability-density distribution or not; only an assessment to what extent. Although sometimes a vapour trail may remain sufficiently concentrated to provide a working criterion of identity, it may not. And in any case, in principle it is not satisfactory, since non-infinitesimal probability-densities are characteristically assigned even to points well outside the band of high concentration. Any point $(x'_1, x'_2, \ldots, x'_n)$ could be correlated with the initial conditions $(x_1, x_2, \ldots, x_n)$ after an interval t, although with some values of $x'_1, x'_2, \ldots, x'_n$, it will be fairly improbable, while with others quite likely. We can no longer say of some values of $x'_1, x'_2, \ldots, x'_n$, that they cannot describe, after t, a system with initial conditions $x_1, x_2, \ldots, x_n$, and of other values of $x'_1, x'_2, \ldots, x'_n$, that they definitely do. We cannot say definitely that a system described at one time definitely is, or definitely is not, the same as a system described at another. Our criterion of identity, or at least of re-identification, has gone.

Two minor points need to be made to ward off possible misunderstandings. The loss of the continuity criterion comes neither with the introduction of densities nor necessarily with that of probabilities, but only with the introduction of the combination of them both. If we were keen on densities pure and simple, we could introduce them by generalising our notion of a causal cord in the following way: instead of having a set of initial conditions which define just one point in $(x_1, x_2, \ldots, x_n)$-space, and which determines just one causal cord, we could have a density-function ranging over $(x_1, x_2, \ldots, x_n)$-space, generating a "causal cable". After any interval, t, every point $(x_1, x_2, \ldots, x_n)$ will be transformed into a point $(x'_1, x'_2, \ldots, x'_n)$; and if we assign to each point $(x'_1, x'_2, \ldots, x'_n)$ the same density as we did to the corresponding point $(x_1, x_2, \ldots, x_n)$, we shall have the density distribution for the space at date t after the date of the initial conditions. We can, equally well, view this not as assigning densities to points in the n-dimensional phase-space, but as assigning densities to causal cords themselves in the $(n + 1)$-dimensional (phase-plus-time)-space. Some with a relatively high density assigned will form the core of the "cable":

others with a low, or no, density will fade into the background. Nevertheless, in spite of our dealing with continuously variable densities instead of one all-or-nothing cord, there is no loss of identity or time-indifference or reversibility. For the transformation we are using from $(x_1, x_2, \ldots, x_n)$-space to $(x'_1, x'_2, \ldots, x'_n)$-space is a *point-to-point* transformation, transforming every point in the one space into one and only one point in the other space. The densities are merely attached to the points like weights. We still go from point to point, and any one set of points is as good as any other. It is only when we are dealing with *probability*-densities that we get trouble, because then we are starting from a point but ending up with a (probability-) density distribution. We do not have a set of point-to-point transformations, each point happening to be assigned a density, but a point-to-density-distribution transformation, and this clearly cannot provide a continuity criterion, nor will it admit of time-indifference or time-reversibility. Probability-densities are not just densities tacked on to a basically discrete, determinate structure, although they presuppose it, since only when the subject has been fixed, and a question asked about it, can a probabilistic answer be given. But for that very reason, although nested within it, they necessarily alter it in a radical fashion, and are themselves entrenched in an entirely different scheme.

Even if we have probabilities, provided they are not probability-densities, we can sometimes keep the continuity criterion. Instead of defining a particular classical system by *one* set of initial conditions, developing along *one* world-line or causal cord, we could have a number of sets of initial conditions, and corresponding causal cords, and assign to each set of initial conditions, or to each causal cord, a definite probability. Then, provided the total probability added up to 1, we should have defined a system in which it did not have to be either true or false of any set of values of the variables at a given time that they satisfied the equations: to a finite number of such sets of values we could ascribe definite, non-infinitesimal probabilities. If we picture it in $(n + 1)$-dimensional phase-space, we should have instead of a single causal cord, a number of causal strands. One could even allow the probability assigned to one causal strand to increase, and another to diminish, so long as none of them increased to unity or decreased to zero: for we should have a sharp discontinuity between points on a causal strand and points off it, and this would give us our criterion of sameness and difference. It is only when the probabilistic magnitude (either a probability or a probability-density) can vary continuously not only along the curve, with time, but off it, with change of values of the variables alone, that we can no longer apply the discrete concepts of 'same' and 'different' in the sense of numerical identity and numerical difference.

Causal cords are time-indifferent. The initial conditions are no different from the conditions at other dates, except that we choose to make them the starting point of our calculations. We could equally well start from the conditions at any other date, with a correspondingly different interval t having elapsed: and in all normal cases we can calculate backwards in time as well as forwards; that is, we can calculate from some subsequent set of conditions what any earlier set of conditions must have been. This is because the initial conditions and the final conditions are on the same logical level. A functional dependence is a mutual dependence (provided the Jacobian is not zero or infinite). The subsequent values of the variables x_1', x_2', ..., x_n', depend, given t, on the initial values of the variables x_1, x_2, ..., x_n; but equally the initial values of the variables depend on their subsequent values. There is a functional relationship between them which can be taken either way. For if any one description of the present state of affairs is true, every other description is false, and therefore every antecedent description from which one of these false descriptions could have been inferred must have been false too. And so, just as all except one of the possible present descriptions is false, all except one of the possible descriptions at any given previous date must be false. Hence, by exhaustion, we can select the remaining one which must be true. Thus we can retrodict just as well as we can predict. Not so with probability, for probabilities are not similarly symmetrical. The probability of $G(f)$, of the propositional function $G(\ )$ in the universe of discourse specified by F, is not the same as, nor related to, the probability of $F(g)$, of the propositional function $F(\ )$ in the universe of discourse specified by G. For whereas the truth of the proposition 'if q then r' enables us to argue not only from the truth of q to the truth of r, but from the falsehood of r to the falsehood of q, the assignment of a probability value α $(0 < \alpha < 1)$ to the propositional function

$F(g)$ does not license any inference whatever from the falsehood of F. As the Churchill example showed,† even if we assign a high probability to the propositional function that a twenty-year-old man who smokes 40 cigarettes a day will die before he is ninety, it does not at all follow that a ninety-year-old man—*i.e.* a man who did *not* die before he was ninety—did not smoke 40 cigarettes a day when he was twenty. Probable propositional functions, unlike true conditional propositions, are one-way-only in the inferences they license, because they are essentially in subject-predicate form and one cannot negate the subject in the way one can negate the predicate.‡ The inverse argument by contraposition and exhaustion, which is available for functional dependences, is not available for probabilistic ones. They are not symmetrical, either as regards inferences, or, therefore, as regards time. And so we cannot expect to have time-reversibility when we are dealing with probabilistic concepts.

Time in the Newtonian philosophy is not only directionless, but date-indifferent: it does not matter which date we calculate from, or which set of conditions we regard as initial conditions. But the disparity between subject and predicate in probability-judgements precludes our being able to make an arbitrary choice of what conditions to regard as initial conditions, in much the same way as it prevents our choosing the direction of argument. The initial conditions of a causal vapour trail are, as it were, the subject; they specify the universe of discourse, and define what it is that we are talking about. They conform to a discrete, two-valued, logic: a particular specification *either* does *or* does not apply; either we are talking about the universe of discourse specified, or we are not. They lay down and define the topic of discourse. The subsequent conditions of a causal vapour trail are, as it were, the predicate, expressed by propositional-functions. They do not lay down the topic of discourse, but rather ask questions about it, questions to which we can give only more-or-less, and not black-or-white, Yes-or-No, answers. Questions are not the same as stipulations. And therefore initial and subsequent conditions cannot exchange *rôles*. A subsequent condition gives us a probability-density for every point throughout the whole of the space, so that for every range or volume, we can say

† See above, Ch. VIII, p. 127.
‡ See above, Ch. IV, pp. 59–60.

what the probability is of the values of the variables falling within that range or volume after an interval t. If we are asked 'Will the values of $x'_1, x'_2, \ldots, x'_n$ lie between a_1 and b_1, a_2 and $b_2, \ldots, a_n$ and b_n after interval t?' we can give, not a Yes-or-No answer, but at least some sort of answer, namely a probabilistic one, for example, 'There is 0·1 probability of their doing so.' But we cannot say we are talking 0·1 about a system with values of x_1, $x_2, \ldots, x_n$, lying between a_1 and b_1, a_2 and $b_2, \ldots, a_n$ and b_n, and 0·9 about something else. We must talk about what we are talking about, and not anything else, or communication breaks down. The subject differs from the predicate not only in not being able to be negated, but in not admitting of degrees.† And so we cannot convert a subsequent condition into an initial condition, except in the special case where the subsequent condition yields a probability of unity: [in that case, provided the question was non-vacuous (*i.e.* the range or volume bounded by a_1 and b_1, a_2 and $b_2, \ldots, a_n$ and b_n, is less than the whole space), since the answer was an unqualified Yes, we could use it to specify a new universe of discourse, a new subject of conversation; and we could then go on to talk about such a universe of discourse, and to pose questions about its subsequent conditions, and offer answers: but only because we are 100% talking about *it*.] In all other cases, where the subsequent condition is expressed in probabilistic terms, it cannnot specify a universe of discourse in which further questions may be asked. Unless a question is answered by an unqualified Yes, it cannot be the starting point of further inquiry.

Although probabilities and probability-densities can vary continuously, universes of discourse cannot. Probabilistic answers can take any value in the closed interval [0, 1]. With probability-densities, though not with probabilities, the questions can vary continuously too. But both with probabilities and even with probability-densities, the subject of discussion—the universe of discourse—cannot vary continuously. This is not to say we cannot ever change the subject—life, or at least speech, would then be very boring—but changes of subject must be definite and sharp. Even within probability theory, we can change from one universe of discourse to another: the unrestricted Conjunction

† See above, Ch. X, p. 175 and note.

Rule gives us the rule for making the change.† But such a change must needs be sharp. We narrow down the universe of all those things that are G to the universe of all those things that are both G and F. The addition of the further specification that things should be F as well as being G is an all-or-none matter. We must either add F to our specification, or not. And if we do, we are talking about a new, although related, subject, and we shall be operating in a new, although related, universe of discourse, in which all assignments of probabilities and probability-densities will have changed discontinuously, although not randomly. So too if we alter the specification of the initial conditions of a vapour-trail we shall discontinuously change the probability-densities throughout the space. If we merely add or subtract some condition, the probability-densities may merely be multiplied or divided by the same factor throughout. But if we alter the universe of discourse in any less simple way, we are likely to alter all probabilities and probability-densities in a more radical fashion. We are talking about something different, so naturally we give different answers, even though the questions are the same. We go on asking 'Does it have values of x_1' lying between a_1 and b_1?' but it is a different *it* we are asking about.

Physicists sometimes talk of the experimenter having to *interfere* with a system in order to observe them. But it is a logical and not an engineering term.‡ To interfere in this sense does not mean to disturb, but to ask a new question of a new subject. In order to measure one physical magnitude we set up one sort of experiment, in order to measure another we set up another experiment. It is not simply that the different experimental set-ups get in each other's way, but rather that they may define different subject-matters. In the two-slit experiment the universe of discourse is an electron-which-has-gone-through-the-two-slitted-screen, and the analogue to the propositional function is the question where it will hit the sensitive plate; and the answer is a probability-density, varying from place to place. If we now try to discover which of the slits some electron goes through, we may well be able to do so, but cannot help changing the subject of the investigation. It will no longer be an electron-which-has-

† See above, Ch. IV, pp. 65–6.
‡ I owe this phrase to J. Bronowski, in his Condon lectures, *Nature and Knowledge*, Eugene, Oregon, 1969, p. 28.

gone-through-the-two-slitted-screen, but an electron-which-has-gone-through-this-specified-slit: and the probability-density distribution in that universe of discourse is no more the same as in the other than the probability of an Englishman having a blue right eye is the same as the probability of an Englishman-with-a-brown-left-eye having a blue right eye.† In each pair of cases we are asking the same question—Where does it hit the sensitive plate? Has he got a blue right eye?—but we are asking the same question of different subjects. It is easy to overlook this change: partly because of the sameness of the question; also, in the case where we are dealing with probability-densities because our usual symbolism does not refer to the universe of discourse explicitly;‡ and in the case of the Englishman because we think that there is some one definite person, Smith, who is there and remains the same, however we describe him.§

In the classical metaphysics one mark of thing-hood, or substance, was independence of what people said or thought or knew about it. Things were there, irrespective of our knowledge of them, each preserving its own identity, each having properties or parameters which we could in principle determine if we chose. Once we had identified a substance by discovering its properties or parameters at any one time, we could re-identify it at other times. And the underlying reason in logic is that what we can say about a thing is in the same logical category as the means whereby we identify and refer to it. We talk about a thing in definitely ascribing to it properties and parameters, and we can use such a definite description to refer to it. But in probability theory we lose parity of esteem between subject and predicate. We no longer need definitely ascribe properties and parameters in talking about a thing, but may do so by halves or quarters or 73·5%s. Our specification of the universe of discourse, however, has to remain definite and all-or-none. We still need definite descriptions to refer with, but no longer obtain them in the natural course of events as alternative descriptions of what has already been described. A classical thing exists out there, and it does not matter whether I refer to it as the thing that had such-and-such properties or parameters at 10 a.m. yesterday, or as the thing that has such-and-

† See above, Ch. IV, pp. 50–3.
‡ See above, Ch. X, pp. 176–7.
§ See above, Ch. VI, pp. 104–8.

such, different, properties or parameters now. For the thing that answered to the one description will answer to the other, and *vice versa*. But when the law of development through time is probabilistic, that which I referred to yesterday cannot be referred to by means of the properties or parameters it has now, because *it* does not *have them*; there is only a probability-density of any particular variable having any particular value. And although if we insist and frame a question by means of a suitable experiment which elicits a definite answer, we can say what the value is now, it is no longer the same *it* in the traditional sense, because the new *it* has definitely the value elicited by experiment whereas the old *it* did not now have any value definitely. It is like having a questionnaire. In the classical view, there was a complete list of questions to which definite answers would be given if the questionnaire was served on a particular thing, *and to which the same answers would in principle be given if the questions were in fact not asked*. In our new metaphysics, apart from the fact that we do not believe that all the questions can be answered at once, we cannot believe that there are potential and definite answers waiting to be given to unasked questions, and these are just the same whether or not the questions are actually asked. So long as the questionnaire is not served, there are only indefinite, probabilistic, answers, and these can be calculated, as it were in the office, from answers previously given. Once served, the questionnaire demands, and gets, a different sort of answers, namely non-probabilistic, definite Yes-or-No, ones. And therefore we can no longer ascribe the same answers to questions unasked as we should obtain if we actually asked them. We are no longer dealing with things which possess properties independently of our determining what they are.

We are not really dealing with things. Or, rather, the word 'thing' has two senses: it may be used as the most general noun for any*thing* we are talking about. In this sense, to be a thing is to be referable to, and sub-atomic entities are things along with Time, Space, Numbers, Qualities, Values, and God. But in the other sense of the word 'thing', in which things have a spatio-temporal location, have spatio-temporal continuity as a criterion of identity, and possess properties or parameters independently of our knowledge of them, the subjects of probabilistic discourse are not things. Although in some respects like Aristotle's primary and secondary substances, they are not exactly like either. Like a secondary substance, a universe of discourse is a species: Englishmen-with-brown-left-eyes, tosses-of-a-coin, corn-grown-with-a-potash-fertiliser-applied, electrons-going-through-a-two-slitted-screen. But unlike secondary substances, qualities, and other universals, the subjects of probabilistic discourse can be individuated in more than one instance without there being any qualitative difference between them. It makes sense to talk of two Englishmen, two electrons-going-through-a-two-slitted-screen, in a way in which it does not make sense to talk of two species of *homo sapiens*,† two hards—where 'hard' is a sortal adjective—or two waters—where 'water' is a bulk noun. Englishmen and electrons are countable in probability theory, and we can calculate the probability of two Englishmen both having blue right eyes or the probability-density distribution for two electrons. For the subject term of a propositional function is not simply a universe of discourse but rather an individual variable ranging in a universe of discourse: it is an *individual* variable, not a predicate variable nor a quantity variable, and therefore it can be counted, although since it is an individual *variable* there is no need for it to be identifiable as any particular individual. It is in this respect that the subjects of probability discourse differ from Aristotle's primary substances. Although they can be articulated, it does not follow that they can be particulated. We can use the indefinite article—an Englishman, another Englishman, some Englishmen, an electron, another electron, some electrons—but we have no assurance that we can go on to particularise an individual enough to be able to identify it uniquely and re-identify it if need be, and so be justified in using the definite article. With Englishmen, tosses of a coin, and fields of corn we have, on other grounds, an assurance that we can, but in quantum mechanics we often have no reason to suppose so but rather the reverse. When we are calculating the probability of getting one head and one tail in two tosses, we distinguish $H(g_1)$ & $\overline{H}(g_2)$ from $\overline{H}(g_1)$ & $H(g_2)$, because g_1 and g_2 can be themselves individuated by time: one is the first toss, the other the second. But when we are considering—to take a crudely simplified example—the probability of two

---

† It makes sense to talk of two *sub*-species of a species. But then there must be some qualitative difference to distinguish the sub-species from each other.

electrons being one in one volume the other in another, we cannot count $H(e_1)$ & $\bar{H}(e_2)$ as a different case from $\bar{H}(e_1)$ & $H(e_2)$, because e_1 cannot be distinguished from e_2. Instead of the Maxwell-Boltzmann statistics which apply to identifiable individuals, we have to apply Bose-Einstein or Fermi-Dirac statistics which work on the assumption that individual instantiations of the type of entity under consideration can be counted but cannot be otherwise identified. In the terminology we assumed at the beginning of Chapter IV,† we should exercise caution before thinking we can name the instances of the individual variable by attaching subscripts to them, $g_1, g_2, \ldots, g_n, \ldots$, although we can number them in the sense of counting them. Another way of expressing this would be to go on allowing the use of numerical subscripts, but with an additional commutative rule to secure that if we have two g's, g_1 and g_2 occurring in an expression, they may be interchanged without its making any difference. Such a commutative rule would distinguish these subjects of probabilistic discourse from traditional primary substances: they would be distinguished from secondary substances in not conforming to any idempotent rule. Thus in Whitehead and Russell's calculus $G(x)$ & $G(x) . \equiv G(x)$; we say no more of a thing if we say it is G and G than we do if we say simply that it is G. But g and g is not the same as g. Two electrons are undoubtedly two, however indistinguishable they are.

The best analogy is money. Electrons are like pennies in the bank. They can be counted. They can be moved from one account to another. But I cannot say—I cannot attach any meaning to the question—which pennies they are. If I lend you fifty pounds to tide you over until your grant arrives, and you pay me back by writing a cheque, I cannot say 'But those are not the same pounds as I lent you'. In the cashless economy where we transact all our business by cheque and credit card, it is a strictly meaningless form of words. At a more primitive level, although not meaningless, it is still unreasonable and absurd. If I borrow a bottle of sherry from you, the demand that I should return to you the selfsame numerically identical bottle is intelligible—but would defeat the whole point of the loan. A bottle of sherry is what the lawyers call "fungible". Its importance, and therefore its existence as a matter of legal concern, depends on its being able to be

† p. 47.

put to a certain use: and anything else which serves the purpose equally well is as good as being the same thing. In the case of money, its whole *raison d'être* is to be fungible, and although different coins regarded as physical objects are numerically distinct, it is conceptually incoherent, and often legally forbidden, to distinguish between them. If we are asked how many ways there are of giving a man one shilling, we naturally do not discriminate between different coins of the same denomination— *e.g.* having specified a shilling, two sixpences, one sixpence and two threepenny bits, we do not then go on to specify the other sixpence and two threepenny bits, *etc.* But we are conscious of a certain measure of artifice. "Things are fungible or not fungible, not in their own nature, but with reference to the terms of the given obligation", said Austin,† and we are easily shaken back into Maxwell-Boltzmann statistics if pressed to consider what coins "really" are. If, however, we accept a thorough-going probabilistic metaphysics, we shall be forced to the opposite conclusion, that things are in their own nature fungible, and it is only by adventitious circumstance or intervention on our part that they acquire any momentarily unique identity. We set up a two-slitted screen in front of a source of electrons. The screen acts, so to speak, as a filter—only electrons which get through the screen get to the other side. If we then put a detecting device—a geiger counter or a photographic plate, say—we can identify electrons by reference to their evident effects—the one that made a click just *then*, the one that darkened the plate just *here*—but we cannot wish on them any continuing identity beyond that of being a screen-traversing-electron which made a click then, or hit the plate here. We cannot say, we cannot even think, which electrons were those that got through the screen and which were those that were stopped, any more than I can say which of my pounds I bought my road fund licence with and which I bought theatre tickets with. The phrases 'The twenty pounds I had to spend on a licence' or 'The twenty-five shillings I spent on theatre tickets' seem to identify sums of money, but they do so only within the context of identification, and not independently of it. I can say that the twenty pounds I spent on a licence set me back

† J. Austin, *Lectures on Jurisprudence*, London, 1885, Vol. II, lecture xlvi, 780. I owe this quotation and much illuminating discussion to Mr. M. J. Lockwood, of Exeter College, Oxford.

a lot, or the twenty-five shillings were completely wasted; but I cannot identify these sums in my bank account apart from the spending of them; and therefore, in particular, cannot envisage my devoting the ticket money to the licence, and some of the licence money to the ticket as a real alternative to what actually happened. In the same way we cannot identify electrons or other sub-atomic particles independently of the context of identification. They have no continuing identity of their own, and many of the questions we Newtonianly want to ask are questions that cannot be asked of them. As London and Bauer put it

le filtre n'amène nullement un objet *individuel* en un état pur nouveau; il ne peut l'amener qu'en état de mélange; c'est ce qui a toujours lieu lorsqu'on couple un système avec un autre. Par contre, nous pouvons évidemment dire que ceux des atomes qui ont passé la fente ont la propriété désirée, et nous pouvons leur attribuer la fonction d'onde du cas pur en question. Mais cela ne se fait pour ainsi dire qu'aux dépens de l'individualité de l'objet, lorsqu'on ne sait pas d'avance *quels* sont les atomes qui ont la propriété en question. Nous pouvons bien attribuer aux objets après la fente la fonction ψ du cas pur, mais nous ne pouvons pas dire *quel objet*, c'est-à-dire quelle variable est l'argument de cette fonction ψ. . . . le filtre seul produit donc bien des cas purs, mais sous une forme absolument *anonyme*.†

It seems strange. Brought up to think of things as being like the material objects we are familiar with in our experience or the Newtonian point-particles we are familiar with in our thoughts, we find it counter-intuitive to think of them as anonymous and lacking each its own continuing identity. But if we follow through the logic of a fundamentally probabilistic view of the world, it becomes less strange, and almost what we should expect. For if it is going to be probabilistic and yet reasonably exact, it must ascribe probabilities to propositional functions, since with propositions we can only estimate their probabilities, and have no way of checking our estimates. If our basic descriptions are in terms of propositional functions, then they are likely to be describing things that are anonymous, fungible, since, from the point of view of a propositional function, any one instantiation of an individual variable is just like any other. There can be only an adventitious difference between g_1 and g_2. Moreover, once we

† F. London et E. Bauer, *La Théorie de l'observation en mécanique Quantique*, Paris, 1939, §13, pp. 47–8.

generalise from causal cords to causal vapour trails, the criterion of identity that served classical physics so well—*viz*. spatio-temporal continuity—ceases to be applicable, because of the disparity between the discrete, definitely referring, subject term, and the continuously varying probabilistic predicate term. And so we should expect to be dealing with anonymous, impermanent, sub-standard substances.

Our basic descriptions of the world have subject terms that are individual variables rather than the names of individual things. We talk in generalities about Englishmen, Englishmen-with-brown-left-eyes, tosses-of-a-coin, corn-grown-with-a-potash-fertiliser-applied, electrons-going-through-a-two-slitted-screen. But they can have, in appropriate circumstances, individual instantiations, this Englishman, this toss of this coin, this field of corn, this scintillation. It must be possible, or we could attach no empirical meaning to our basic descriptions at all. The Demiurge deals in forms: but we want clear and distinct ideas. In order that the language of forms shall be empirical, we must sometimes be able to put it to the test and extract definite, unequivocal answers. And we can. With the passage of time, generalities are particularised into particular instances with definite details. The Englishman turns out to be Smith: the coin comes down heads: the field of corn bears an hundredfold: the electron went just here. Time is the passage from potentiality to actuality: sometimes with our intervention, sometimes without, what had been on the cards either happens or does not happen, and becomes a definite thing which definitely does or definitely does not possess properties and parameters. The Demiurge is continually being forced to declare his suit. Were it not so, knowledge would be impossible, for his forms, although mathematically defined, are cloudy, and tend to grow cloudier. It is a necessary condition of probabilistic judgements being empirical statements about the world, that there should be occasions when what they guardedly predict either happens or does not happen.† But each time, the situation is radically altered, and we cannot help talking thereafter about the new situation, not the old. New things are constantly happening: old knowledge needs constant refurbishing. The past is only a partial pointer to the future. We can, but only guardedly, guide our expectations of what is to come: we can make predictions,

† See above, Ch. V, p. 94.

but only in general terms and not infallibly, for the future is not yet particularised, and is not fixed or laid up in the past, but is always and inherently open.

The metaphysical foundations of a probabilistic science need more thought. It is enough for the present to show how great are the implications of accepting objective probabilities as part of our ultimate description of the world. Not only does determinism go, but also the Newtonian views that the direction of time is not physically significant, and that a description at any one date is as good as a description at any other. And with this, our basic and cherished concept of *haecceitas* or primary substance.

APPENDIX I

Alternative Proof of Bernoulli's Theorem

(SEE CHAPTER V, pp. 79–83)

TCHEBYCHEV'S Inequality can be used to prove Bernoulli's Theorem, once we have calculated the mean and the standard deviation of the Binomial Distribution. Essentially, the strategy is that of the bear-hug. Having fixed the scale in the mathematically relevant sense, we use Tchebychev's Inequality to establish a hold over the terms not belonging to the central peak, and then, by increasing n, show that the contribution from these terms diminishes indefinitely, and thus we squeeze the mountain into pre-assigned limits.

What is the relevant scale? We are interested in the terms of the Binomial Distribution in as much as the frequency they represent diverges from the probability of the propositional function itself. It might seem that we should be interested in the magnitude $\alpha - \frac{r}{n}$ as a measure of the divergence of each frequency from α. But this is wrong. If we calculate—and for other reasons it is worth making the calculation—the sum total of all the terms of the form

$$\left(\alpha - \frac{r}{n}\right) \times {}^nT_r$$

we shall find that it is zero. To see this we calculate

$$\sum_{r=0}^{n} \left(\frac{r}{n} \times {}^nT_r\right)$$

noting first that this must be the same as

$$\sum_{r=1}^{n} \left(\frac{r}{n} \times {}^nT_r\right)$$

since the first term,

$$\left(\frac{0}{n} \times {}^nT_0\right)$$

is zero. Each of the other terms, where $r \neq 0$, is of the form

$$\frac{r \cdot n!}{n \cdot r! (n-r)!} \cdot \alpha^r (1-\alpha)^{n-r}$$

which reduces to

$$\frac{(n-1)!}{(r-1)!\,(n-r)!}\,\alpha^r(1-\alpha)^{n-r}$$

and can be rewritten

$$\frac{\alpha\,(n-1)!}{(r-1)!\,(n-r)!}\,\alpha^{r-1}(1-\alpha)^{(n-1)-(r-1)}$$

Hence the total sum is

$$\sum_{r=1}^{n}\frac{\alpha\,(n-1)!}{(r-1)!\,(n-r)!}\,\alpha^{r-1}(1-\alpha)^{(n-1)-(r-1)}$$

$$=\alpha\times\sum_{r=1}^{n}\frac{(n-1)!}{(r-1)!\,(n-r)!}\,\alpha^{r-1}(1-\alpha)^{(n-1)-(r-1)}.$$

In this new sum we have exactly the same form as we had for $\sum\,({}^{n}T_r)$ except that n has been replaced by $n-1$, and r by $r-1$. If we write s for $r-1$, we can express the new sum as $\sum_{s=0}^{n-1}{}^{n-1}T_s$. Now $\sum_{s=0}^{n-1}{}^{n-1}T_s$ must be 1, for it is simply the Binomial expansion of $(\alpha+(1-\alpha))^{n-1}=1^{n-1}=1$. It follows that the original sum

$$\sum_{r=0}^{n}\left[\frac{r}{n}\times{}^{n}T_r\right]=\alpha\times1=\alpha$$

and therefore

$$\sum_{r=0}^{n}\left[\left(\alpha-\frac{r}{n}\right)\times{}^{n}T_r\right]=\alpha\times\sum_{r=0}^{n}{}^{n}T_r-\sum_{r=0}^{n}\left[\frac{r}{n}\times{}^{n}T_r\right]=\alpha\times1-\alpha=0,$$

arguing again that $\sum_{r=0}^{n}{}^{n}T_r=1$.

If we reflect we can see why we have reached this surprising result. In summing the contributions made by the different values of r, the contribution made by those r's which are greater than $n\alpha$ have been exactly cancelled out by those which are less, and which have therefore been accorded a negative value. What we have shown is that the *average*—or better, the *mean*†—value of the original distribution is α. α is, so to speak, its centre of gravity. It is something we had rather assumed. It is just as well that we have proved it. But it is not what we want when we are discussing the degree to which the Binomial Distribution diverges from its mean value α.

The trouble arose from some of the contributions being negative, since $(\alpha-r/n)$ must sometimes take negative values. The simplest way to get rid of negative numbers is to square them. $(\alpha-r/n)$ is sometimes negative but $(\alpha-r/n)^2$ never is. Just as in Euclidean geometry we base the essentially non-negative concept of distance on

† See above, Ch. IX, pp. 166–7.

Pythagoras' Theorem connecting the squares of distances, so we measure the way in which the terms deviate from the mean by multiplying each one by $(\alpha - r/n)^2$. We calculate, so to speak, not the centre of gravity, but the moment of inertia. It is clear that this is a fair procedure, in as much by taking squares we emphasize much more the contribution of terms a long way from the peak, and minimise the contribution of those very close to the peak. If, even so, the contribution of those at any significant distance from the peak becomes negligible for sufficiently large n, we shall have achieved our squeeze.

We therefore calculate

$$\sum_{r=0}^{n} \left[\left(\alpha - \frac{r}{n} \right)^2 \times {}^nT_r \right].$$

Expanding the square bracket, and taking out constant factors, we have

$$\alpha^2 \times \sum_{r=0}^{n} ({}^nT_r) - 2\alpha \times \sum_{r=0}^{n} \left(\frac{r}{n} \times {}^nT_r \right) + \sum_{r=0}^{n} \left(\frac{r^2}{n^2} \times {}^nT_r \right).$$

The first of these sums is again a simple Binomial Distribution with a total of 1; the second is the one we have already calculated and found to be equal to α. The whole sum therefore becomes

$$\alpha^2 - 2\alpha^2 + \sum_{r=0}^{n} \left(\frac{r^2}{n^2} \times {}^nT_r \right) = \sum_{r=0}^{n} \left(\frac{r^2}{n^2} \times {}^nT_r \right) - \alpha^2.$$

We therefore need to calculate

$$\sum_{r=0}^{n} \left(\frac{r^2}{n^2} \times {}^nT_r \right).$$

As before, we simplify;

$$\frac{r^2}{n^2} \times {}^nT_r = \frac{r^2 n! \, \alpha^r (1 - \alpha)^{n-r}}{n^2 r! \, (n - r)!} = \frac{r(n - 1)! \, \alpha^r (1 - \alpha)^{n-r}}{n(r - 1)! \, (n - r)}$$

$$= \frac{(r - 1 + 1)(n - 1)! \, \alpha^r (1 - \alpha)^{n-r}}{n(r - 1)! \, (n - r)}$$

$$= \frac{\alpha}{n} \left[\frac{(r - 1)(n - 1)! \, \alpha^{r-1}(1 - \alpha)^{n-r}}{(r - 1)! \, (n - r)!} \right.$$

$$\left. + \frac{(n - 1)! \, \alpha^{r-1}(1 - \alpha)^{n-r}}{(r - 1)! \, (n - r)!} \right]$$

$$= \frac{\alpha}{n} \left[\frac{\alpha(n - 1)(n - 2)! \, \alpha^{r-2}(1 - \alpha)^{(n-2)-(r-2)}}{(r - 2)! \, [(n - 2) - (r - 2)]!} \right.$$

$$\left. + \frac{(n - 1)! \, \alpha^{r-1}(1 - \alpha)^{(n-1)-(r-1)}}{(r - 1)! \, [(n - 1) - (r - 1)]!} \right].$$

When we take the sum, the second terms, as before, yield a sum total of 1; the first term has a constant factor $\alpha\,(n-1)$, and so its sum is $\alpha\,(n-1)\sum\,(^{n-2}T_{r-2})$ which by yet another application of the same argument comes to $\alpha\,(n-1)\times 1$, *i.e.* $\alpha\,(n-1)$. Hence the sum

$$\sum_{r=0}^{n}\left(\frac{r^2}{n^2}\times {}^nT_r\right)=\frac{\alpha}{n}\left(\alpha\,(n-1)+1\right)=\alpha^2+\frac{\alpha\,(1-\alpha)}{n}$$

and thus the total sum

$$\sum_{r=0}^{n}\left(\left(\alpha-\frac{r}{n}\right)^2\times {}^nT_r\right)=\sum_{r=0}^{n}\left(\frac{r^2}{n^2}\times {}^nT_r\right)-\alpha^2=\frac{\alpha\,(1-\alpha)}{n}.$$

The important thing about this result is that $[\alpha(1-\alpha)]/n\to 0$ as $n\to\infty$. This enables us to use Tchebychev's Inequality. Consider all those terms that are more then δ away from the middle; all those terms, that is, for which $|\alpha-r/n|>\delta$, *i.e.* $(\alpha-r/n)^2>\delta^2$. Their contribution must be less, *a fortiori*, than $\sum_{r=0}^{n}((\alpha-r/n)^2\times {}^nT_r)$, which we have just calculated to be $[\alpha(1-\alpha)]/n$; moreover the sum of $(\alpha-r/n)^2\times {}^nT_r$ for all the distant terms must be larger than the sum $\delta^2\times {}^nT_r$ for those same terms, since these terms are just those for which $(\alpha-r/n)^2$ is greater than δ^2. Now the sum of $\delta^2\times {}^nT_r$ for the distant terms is, since δ^2 is a constant, $\delta^2\times$ [the sum of ^nT_r for the distant terms]. And this is less than $[\alpha(1-\alpha)]/n$. Therefore the sum of ^nT_r for the distant terms is $[\alpha(1-\alpha)]/n\delta^2$ and hence, however small you have chosen δ and ϵ, provided only that they are positive and not zero, I can calculate an N large enough to ensure that, for every $n>N$, the sum of the distant terms is less than ϵ; for so long as I take n greater than $[\alpha(1-\alpha)]/\epsilon\delta^2$, I can be sure that the unacceptably distant—according to your criterion—part of the distribution will constitute only a negligible—according again to your criterion—part of the whole. Or, to put it the other way round, however closely you define the middle and however stringently you define what is negligible, I can find a number beyond which all save a negligible part of the distribution will be in the middle.

APPENDIX II

Summary of Theories of Probabilities

FOR the benefit of those who have to write essays I give brief notes about each of the six theories currently espoused.
I. J. GOOD, "Kinds of Probability", *Science*, **129**, 1959, pp. 443–7, gives a useful *résumé*.

FREQUENCY THEORY

Chapter V, pp. 95–100, Chapter VIII, pp. 140–1

Chief proponents: Venn, von Mises, Reichenbach.
Recommended source: RICHARD VON MISES, *Probability, Statistics and Truth*, 2nd English ed., London, 1957.

Merits   (i) squares with intuitive sense as expressed by 'in the long run' 'in three cases out of four',

      (ii) squares with statistical practice.

Demerits (i) cannot account for the probabilities ascribed to singular propositions,

      (ii) cannot give any clear account of 'in the long run' 'limiting frequency' 'collective',

     (iii) confuses meaning with criteria of use,

     (iv) open to detailed objection on the score of ordinary probability theory,

     (v) otiose, in view of Bernoulli's Theorem.

Criticized: J. M. KEYNES, *A Treatise on Probability*, London, 1921, Ch. VIII, pp. 92ff.

      W. C. KNEALE, *Probability and Induction*, Oxford, 1949, §33, pp. 161ff.

LOGICAL RELATION THEORY

Chapter IV, pp. 49–56

Chief proponents: W. E. Johnson, J. M. Keynes, H. Jeffreys.

Recommended source: HAROLD JEFFREYS, *Theory of Probability*, 3rd ed., Oxford, 1961, Ch. I.

Merits   (i) can account for probabilities being ascribed to singular propositions,

(ii) formalises our locution in which we talk of *arguments* (rather than *propositions*) being probable,

(iii) attempts to formalise role of evidence in probability statements.

Demerits (i) easily leads to subjectivism,

(ii) fails to allow for the fact that we more naturally ascribe probabilities to propositions rather than arguments,

(iii) confuses *evidence* with *description* or *specification*.

Criticized: S. E. TOULMIN, *The Uses of Argument*, Cambridge, 1958, Ch. II, pp. 79–83.

EQUIPROBABILITY THEORY

Chapter VII

Chief proponents: Pascal, most Eighteenth Century writers, Laplace, Kneale.

Recommended source: LAPLACE, *Essai Philosophique sur les probabilités*, 3rd ed, Paris, 1816), tr. F. W. Truscott and F. L. Emory, New York, 1902, Chs. II and III.

Merits   (i) intelligible,

(ii) applies satisfactorily to games of chance and in some scientific contexts.

Demerits (i) fundamental assumption (of Equiprobability) not adequately justified,

(ii) can be made to yield inconsistent results where the Equiprobability assumption is unclear (pp. 117–18),

(iii) inapplicable in many contexts—in biological and social sciences,

(iv) provides no basis for the theory of statistics.

Criticized: J. M. KEYNES, *A Treatise on Probability*, London, 1921, Ch. IV, pp. 41ff.

AXIOMATIC APPROACH

Chapter III, pp. 26–9, Chapter IV, p. 65

Chief proponents: Kolmogorov, Cramér.

Recommended source: A. N. KOLMOGOROV, *Foundations of Probability Theory*, English tr., Chelsea, New York, 1956; or H. CRAMÉR,

Mathematical Methods of Statistics, Princeton, N.J., 1946, §§13.4–
13.5, pp. 145–51, or *Elements of Probability Theory*, New York, 1955,
Part I, Chs. 1–3, pp. 1–39.

Merits   (i) mathematically rigorous,

(ii) enables highly sophisticated refinements to be developed.

Demerits   (i) uninterpreted. All philosophical problems about its
application undiscussed and unresolved.

SUBJECTIVE APPROACH (overlaps with Logical Relation theory)

Chapter II, pp. 12–22, Chapter IV, pp. 52–6,

Chapter VIII, p. 141

Chief proponents: Ramsay, Savage, de Finetti, Good.

Recommended source: F. P. RAMSAY, *The Foundations of Mathematics*, London, 1931, pp. 166–84.

Merits   (i) based on the betting propensity of the British people,

(ii) erects theory of probability on extremely slender basis.

Demerits   (i) assumes that people who bet are rational,

(ii) cannot account for reasonable, intersubjective, assessment of probabilities.

Criticized: R. CARNAP, *Logical Foundations of Probability*, London,
1951, §11, pp. 41–2.

INFORMAL APPROACH

Chapter I

Chief proponent: Toulmin.

Recommended source: S. E. TOULMIN, *The Uses of Argument*, Cambridge, 1958, Ch. II.

Merits   (i) gives reasonably good account of the actual use of the
concept in ordinary language,

(ii) explains why we should be guided by probabilities,

(iii) says nothing false or silly.

Demerits   (i) does not take us very far; in particular, gives no justification of the calculus of probabilities.

Criticized: NEIL COOPER, "The Concept of Probability", *British
Journal for the Philosophy of Science*, 1965.

INDEX

Less important references are given after more important ones. The most important ones are given in bold type.

random, randomness, **112–25**, 46, 98–9, 168
Reason, Principle of Insufficient, 110–12
refer, referring, reference, 46–9, 50–2, 107, 108, 187–8, 190–1, 200–8
referentially opaque, 107
Reichenbach, H., 95n., 213
Relativity Theory, 101–2
Roman Catholics, 169–70
Russell, B., 48, 96, 100, 127, 204

sample, **170–2**
semantics, 24, 72, 85
set theory, 27–8, 86–7
Shakespeare, 88–9, 139–40
significance levels, **134–9**, 157, 159, 172
singular propositions, **101–8**, 96–8
Smiley, T. J., **57–9**, 48, 127
Smith, B. Babington, 121
Smith (Jeffrey's friend), **50**, 51, 52, 97, 104, 107, 188, 201, 207
smoking, and cancer, 8–9, 51, 163, 96–7, 127, 198
Snow, Lord, 164
sorts, many-sorted logic, 59, 202–8
Standard Deviation, **166–7**, 136–7, 211–12, 179
statistic, sufficient, 133n.
statistics, **163–73**, viii; Maxwell-Boltzmann, 204, 205; Bose-Einstein 204; Fermi-Dirac, 204
Stuart, A., 121n.
Stirling's formula, 79n., 180
subject, **59–60**, **197–208**, 47, 51–2, 187–91
Subjective Theory of Probability, **13–21**, **52–6**, **215**, 192–3
Subjectivism, Subjectivist, **13–21**, **52–6**, 12, 192–3

substance, 108, 192–7, 59–60, 201–8
subtraction, **26**, 120
Swedes, 169–70
symmetry, 23, 44, 109–12, 172, 198
syntax 24, 72, 85

Tchebychev, 209–12, 79n.
teleological, 123–4
thermodynamics, 190, 191, 193
things, **59–60**, 108, 201–8
time, **197–202**, 104–8, 192–5, 207, 208
Toulmin, S. E., 2, 15, 16, 215
True, truth, **1–2**, **32**, 3–9, 14, 15, 16, 18, 20, 21, 23–4, 28, 34, 35–6, 38–9, 48, 65, 71, 84–5, 94, 146, 187–8, 197
truth-value, **21**, 19, 25, 47, 48, 105–6

unit element, **32**, 37
universal, universality, universal-isability, 96–8, 101–3, 115, 4–5
universal element, **32**
universe of discourse, **50**, **56–67**, **106–8**, **197–8**, **200–8**, 190–1, 192, 47, 101, 149–53
Urmson, J. O., 14n.

variable, **46–9**, **59–60**, 50–6, 95, 203–4
Venn, J., 95n., 213

Walker, A. G., 36n.
Whitehead, A. N., 48, 127, 204
Whitrow, G. J., 36n.
Whitworth, W. A., 19n., 109
Wright, G. H. von, 85n.
Wykehamists, Old, 171